BIOLOGICAL CONFINEMENT

OF GENETICALLY ENGINEERED ORGANISMS

Committee on Biological Confinement of Genetically Engineered Organisms

Board on Agriculture and Natural Resources

Board on Life Sciences

Division on Earth and Life Studies

NATIONAL RESEARCH COUNCIL
OF THE NATIONAL ACADEMIES

THE NATIONAL ACADEMIES PRESS
Washington, DC
www.nap.edu

THE NATIONAL ACADEMIES PRESS   500 Fifth Street, N.W.   Washington, DC 20001

NOTICE: The project that is the subject of this report was approved by the Govern-ing Board of the National Research Council, whose members are drawn from the councils of the National Academy of Sciences, the National Academy of Engineer-ing, and the Institute of Medicine. The members of the committee responsible for the report were chosen for their special competences and with regard for appropriate balance.

This study was supported by Agreement No. 59-0790-1-182 between the National Academy of Sciences and the U.S. Department of Agriculture. Any opinions, find-ings, conclusions, or recommendations expressed in this publication are those of the author(s) and do not necessarily reflect the views of the organizations or agencies that provided support for the project.

Library of Congress Cataloging-in-Publication Data

Biological confinement of genetically engineered organisms / Committee on Biologi-cal Confinement of Genetically Engineered Organisms, Board on Agriculture and Natural Resources, Board on Life Sciences, Division on Earth and Life Studies.
    p. cm.
  Includes bibliographical references and index.
  ISBN 0-309-09085-7 (hardcover)—ISBN 0-309-52778-3 (pdf)
  1. Transgenic organisms—Safety measures. 2. Confinement farms. 3. Agricultural biotechnology. 4. Infertility in animals. 5. Transgenic organisms—Risk assessment. I. National Research Council (U.S.). Committee on Biological Confinement of Ge-netically Engineered Organisms.
  QH442.6.B54 2004
  577'.18—dc22
                        2004004051

Additional copies of this report are available from the National Academies Press, 500 Fifth Street, N.W., Lockbox 285, Washington, DC 20055; (800) 624-6242 or (202) 334-3313 (in the Washington metropolitan area); Internet, http://www.nap.edu

THE NATIONAL ACADEMIES
Advisers to the Nation on Science, Engineering, and Medicine

The **National Academy of Sciences** is a private, nonprofit, self-perpetuating society of distinguished scholars engaged in scientific and engineering research, dedicated to the furtherance of science and technology and to their use for the general welfare. Upon the authority of the charter granted to it by the Congress in 1863, the Academy has a mandate that requires it to advise the federal government on scientific and technical matters. Dr. Bruce M. Alberts is president of the National Academy of Sciences.

The **National Academy of Engineering** was established in 1964, under the charter of the National Academy of Sciences, as a parallel organization of outstanding engineers. It is autonomous in its administration and in the selection of its members, sharing with the National Academy of Sciences the responsibility for advising the federal government. The National Academy of Engineering also sponsors engineering programs aimed at meeting national needs, encourages education and research, and recognizes the superior achievements of engineers. Dr. Wm. A. Wulf is president of the National Academy of Engineering.

The **Institute of Medicine** was established in 1970 by the National Academy of Sciences to secure the services of eminent members of appropriate professions in the examination of policy matters pertaining to the health of the public. The Institute acts under the responsibility given to the National Academy of Sciences by its congressional charter to be an adviser to the federal government and, upon its own initiative, to identify issues of medical care, research, and education. Dr. Harvey V. Fineberg is president of the Institute of Medicine.

The **National Research Council** was organized by the National Academy of Sciences in 1916 to associate the broad community of science and technology with the Academy's purposes of furthering knowledge and advising the federal government. Functioning in accordance with general policies determined by the Academy, the Council has become the principal operating agency of both the National Academy of Sciences and the National Academy of Engineering in providing services to the government, the public, and the scientific and engineering communities. The Council is administered jointly by both Academies and the Institute of Medicine. Dr. Bruce M. Alberts and Dr. Wm. A. Wulf are chair and vice chair, respectively, of the National Research Council.

www.national-academies.org

Acknowledgments

This report represents the integrated efforts of many individuals. The committee thanks all those who shared their insight and knowledge to bring the document to fruition. We also thank all those who provided information at our public meetings and who participated in our public sessions.

During the course of its deliberations, the committee sought assistance from several people who gave generously of their time to provide advice and information that were considered in its deliberations. Special thanks are due the following:

WILLY DE GREEF, Syngenta Seeds, Basel, Switzerland
CATHLEEN ENRIGHT, APHIS / USDA, Riverdale, Maryland
PHILIP J. EPPARD, Monsanto Protein Technologies, St. Louis, Missouri
PAL MALIGA, Rutgers University, Piscataway, New Jersey
MICHAEL H. PAULY, Epicyte Pharmaceutical, Inc., San Diego, California
JANE RISSLER, Union of Concerned Scientists, Washington, DC
MICHAEL SCHECHTMAN, USDA, Washington, DC
ANTHONY M. SHELTON, Cornell University, Geneva, New York
STEVEN H. STRAUSS, Oregon State University, Corvallis, Oregon

The committee is grateful to members of the National Research Council (NRC) staff who worked diligently to maintain progress and quality in its work. We also would like to thank Robert McDonald, Melissa Brandt, Sarah De Belen, Tazeen Hasan, and Rene Milet for their research assistance.

This report has been reviewed in draft form by individuals chosen for their diverse perspectives and technical expertise, in accordance with procedures approved by the National Research Council's Report Review Committee. The purpose of this independent review is to provide candid and critical comments that will assist the institution in making its published report as sound as possible and to ensure that the report meets institutional standards for objectivity, evidence, and responsiveness to the study charge. The review comments and draft manuscript remain confidential to protect the integrity of the deliberative process. We wish to thank the following individuals for their review of this report:

DAVID ADELMAN, University of Arizona College of Law, Tucson, Arizona
KLAUS AMMANN, University of Bern, Bern, Switzerland
R. JEFFREY BURKHARDT, University of Florida, Gainesville, Florida
JEFFREY DANGL, The University of North Carolina, Chapel Hill, North Carolina
DONALD N. DUVICK, Iowa State University, Ames, Iowa
VIRGINIA S. HINSHAW, University of California, Davis, California
CALESTOUS JUMA, Harvard University, Cambridge, Massachusetts
STEVEN E. LINDOW, University of California, Berkeley, California
TERRY MEDLEY, DuPont Agriculture and Nutrition, Wilmington, Delaware
WILLIAM MUIR, Purdue University, West Lafayette, Indiana
CHRISTOPHER SOMERVILLE, Stanford University, Stanford, California
STEVEN STRAUSS, Oregon State University, Corvallis, Oregon
GREG TRAXLER, Auburn University, Auburn, Alabama

Although the reviewers listed above have provided many constructive comments and suggestions, they were not asked to endorse the conclusions or recommendations nor did they see the final draft of the report before its release. The review of this report was overseen by Dr. Robert A. Frosch, Harvard University, Cambridge, Massachusetts, and Dr. Fred Gould, North Carolina State University, Raleigh. Appointed by the National Research Council, they were responsible for making certain that an independent examination of this report was carried out in accordance with institutional procedures and that all review comments were carefully considered. Responsibility for the final content of this report rests entirely with the authoring committee and the institution.

Preface

Genetically engineered microbes have been used commercially for many years to make products useful to humans. Production is confined to vessels that are sterilized between batches. There has been little concern about these genetically engineered (GE) microbes escaping into the wild and doing damage, in part because they are confined physically and in part because they are weakened by foreign genetic material that makes them unlikely to survive in the wild.

But 10 years ago genetically engineered crop plants were introduced into field environments, a situation quite different from having GE microbes in fermentors. This outplanting of GE plants raised, and continues to raise, public and scientific concerns on the potential consequences of escape of genetically engineered organisms (GEOs) and their associated transgenes into natural and managed ecosystems. The acreage planted with GE crop plants has steadily and rapidly increased, as have the number of types of GE plants. GE fish and other animals have now been developed, some with remarkable potentials. This increased use and development of new GEOs obviously reflects the fact that GEOs can have substantial advantages over their progenitors. Agricultural biotechnology has enormous potential to better the human condition. However, concern about the possible risks posed by some GEOs has led to questions about the regulation they receive, and has increased interest in assuring that certain GEOs are confined. While no serious consequences have ensued because of a failure of GEO confinement, with the growing diversity and number of GEOs being developed and released, the potential for unwanted consequences increases, and scientists can envision undesirable scenarios.

Confinement can be accomplished not only by physical but also by biological means. Several examples of long-used biological confinement methods (used with non-GEOs) are described in this report. Genetic engineering, however, makes new biological confinement strategies possible. We refer to all biological confinement methods, whether genetically engineered or not, as *bioconfinement* methods.

The NRC convened a committee of 12 members from a variety of complementary specialties; the reader is urged to read the brief biographical sketches. Members were selected who could not only cover the different aspects of bioconfinement, but who could also assure a flexible view of the committee's charge and provide overall and realistic balance.

Our task was challenging. Bioconfinement of GEOs is in its infancy as a focused science. But the science is fast-moving, rapidly evolving. We found ourselves dealing with a lack of published data on many of the existing and potential methods. Entirely new methods of bioconfinement were announced by the scientific community while we worked on the report. Consequently, we did not focus exclusively on methods in commercial use but tried to anticipate future developments. An exhaustive literature search was not part of our charge; only a few illustrative examples were provided for each confinement approach. Our statement of task of six questions (see the Executive Summary and Chapter 1) limited us further because effective bioconfinement will require more than sound science. It will require safe practices and commitment by those who design and develop the GEOs; effective regulatory oversight; a public commitment to investing in this technology; a high level of public confidence and acceptance; effective communication between stakeholders; respect for regional and cultural differences in values, experience, ethics; recognition of the international dimensions; and more.

We met four times, for a total of about eight days. Most of the research and writing was done individually between meetings, so that the meetings could be devoted to critique and improvement. We began with presentations by and discussions with the sponsoring agency and outside experts from industry, academia, and nongovernmental organizations (NGOs). We then prepared a report outline which was continuously revised, and wrote the report via numerous iterations. This report is a consensus document. Each committee member had the opportunity to question and modify the content of each paragraph. We all learned in this process. The procedure resulted in improvements in every section. In-depth reviews by outside experts from academia and industry resulted in further improvements in the report's clarity, balance, and presentation.

Chapter 1 provides an introduction to the subject, gives the scope of the study, provides a brief history of GEOs and their confinement, and gives an introduction into ethical and other social considerations. The committee recognized that many GEOs will not warrant confinement. We therefore addressed the issue of when and why bioconfinement might be necessary and described briefly some of the possible undesirable consequences of escape of some GEOs (Chapter 2). Chapters 3, 4, and 5 detail what is presently known about bioconfinement methodologies for plants, animals, and microbes (and fungi and viruses), respectively. Chapter 6 summarizes the biological and operational considerations for bioconfinement and points to some important research needs.

Serving as chair of this committee was a most interesting and enlightening experience for me personally. Bioconfinement is not my field, nor is genetic engineering, so I was afforded significant learning opportunities. At the outset I was surprised at the breadth of subject matter that "bioconfinement of GEOs" invites. This is especially true since we considered plants, animals and microbes. It is my hope, that we have written a report that not only will be valuable to the sponsoring agency and other stakeholders, but also one that reflects the committee's concerted effort to adequately characterize a rapidly evolving and complex subject in a National Academies' report. I also hope that our efforts contribute much to the ongoing discussion about biotechnology and the opportunities to utilize the biological tools available to minimize the concerns and risks surrounding the release of GEOs into the environment. Most of all, I hope readers enjoy what we have prepared.

I want to thank the committee members sincerely for their participation and hard work; they are all busy people, and NRC work is *pro bono*. We all thank our NRC study director, Dr. Kim Waddell, for his excellent leadership, and we thank his staff of Julie Coffin, Michael Kisielewski, Cindy Lochhead, Peter Rodgers, and Donna Wilkinson.

T. Kent Kirk, Chair
Committee on Biological Confinement
of Genetically Engineered Organisms

Contents

EXECUTIVE SUMMARY   1
    Rationale for Bioconfinement, 3
    Methods of Bioconfinement, 4
    Ensuring Bioconfinement Efficacy, 6
    Detecting and Mitigating Bioconfinement Failure, 10
    Ecological Consequences of Large-Scale Use of Bioconfinement, 11
    Conclusions, 12

1   INTRODUCTION   14
    What Are Genetically Engineered Organisms?, 14
    What is Bioconfinement?, 15
    Other Confinement Methods,16
    Scope of the Report, 17
    International Aspects, 19
    History of Confinement, 19
    Social Acceptability of Bioconfinement Methods, 25

2   WHEN AND WHY TO CONSIDER BIOCONFINEMENT   29
    Introduction, 29
    What is Risk?, 30
    Concerns, 35
    Effects on Nontarget Species, 52
    Delaying the Evolution of Resistance, 52
    Food Safety and Other Issues, 53
    When and Why to Consider Bioconfinement: The Need for
        Preventive Actions, 53

How Much Confinement Is Enough?, 54
Need for Bioconfinement, 55
Predicting the Consequences of Failure, 56
Who Decides, 58

3   BIOCONFINEMENT OF PLANTS   65
    Methods of Bioconfinement, 65
    Genetically Engineered Trees, 98
    Transgenic Grasses, 115
    Transgenic Algae, 121
    Effectiveness at Different Spatial and Temporal Scales, 122
    Monitoring and Managing Confinement Failure, 124

4   BIOCONFINEMENT OF ANIMALS: FISH, SHELLFISH,
    AND INSECTS   130
    Bioconfinement of Fish and Shellfish, 132
    Bioconfinement of Insects, 153

5   BIOCONFINEMENT OF VIRUSES, BACTERIA, AND
    OTHER MICROBES   159
    Introduction, 159
    Potential Effects or Concerns, and Need for Bioconfinement in
        Viruses, Fungi, and Bacteria, 160
    Bioconfinement of Bacteria, Viruses, and Fungi, 169

6   BIOLOGICAL AND OPERATIONAL CONSIDERATIONS
    FOR BIOCONFINEMENT   180
    What Biology Tells Us about Confinement and
        Bioconfinement, 180
    Execution of Confinement, 185
    International Aspects, 193
    Bioconfinement Failure, 194
    Looking to the Future: Strategic Public Investment in
        Bioconfinement Research, 195

REFERENCES   199

ABOUT THE AUTHORS   235

BOARD ON AGRICULTURE AND NATURAL RESOURCES
PUBLICATIONS   241

INDEX   245

Tables, Figures, and Boxes

TABLES

2-1 Systematic Risk Assessment and Management, 33
2-2 Genetically Engineered Organisms, 39
3-1 Bioconfinement Methods in Plants, 66
3-2 Genetically Engineered Woody Plants, Permits Approved by APHIS
    for Field Tests in the United States, 1989–2003, 99
3-3 Genetically Engineered Turfgrasses, Permits Approved by APHIS for
    Field Tests in the United States, 1993–2003, 116
4-1 Genetic Bioconfinement Strategies for Fish, 147
4-2 Insects Subjected to the Sterile Insect Technique, 155

FIGURES

2-1 A Risk Assessment Matrix, 32
3-1 Proposed Transgenic Bioconfinement Methods in Plants, 73
3-2 Repressible Seed—Lethal Bioconfinement, 86
3-3 A Wild Hybrid, *F. arundinacea* and *L. multiflorum* Lam, 119
4-1 Normal Steps in Gamete Fertilization and Early Cell Division, 134
4-2 Production Cycle for All-Female Lines of Fish in Species with an XY
    Sex Determination System, 143

BOXES

2-1   Confinement Failure: StarLink Corn, 34
3-1   Stability of Transgenic Confinement, 102
3-2   When Will Bioconfinement be Necessary for Trees?, 105
3-3   Turfgrass Might be Difficult to Confine, 120
4-1   Proposed Bioconfinement of Transgenic Atlantic Salmon, 137
5-1 χ1776, 172

BIOLOGICAL CONFINEMENT
OF GENETICALLY ENGINEERED ORGANISMS

Executive Summary

Since genetically engineered organisms (GEOs) were introduced into the environment nearly 20 years ago, questions have been raised about the consequences of the escape of those organisms and their engineered genetic material—transgenes—into natural and managed ecosystems. Ecological research has shown that some GEOs are viable in natural ecosystems and can cross with wild relatives. There also are instances in which transgenes from one domesticated variety can move to others. As a result, there is interest in developing methods to confine certain GEOs and their transgenes to specifically designated release settings. Many confinement methods, including induced sterilization and other methods, are biological in nature, whereas others rely on physical restrictions such as greenhouses or aquaculture pens. This report refers to these biological methods as *bioconfinement*. Although bioconfinement of GEOs is still largely in the conceptual and experimental stages of development, some methods already have been applied to nonengineered organisms.

The primary mechanism in the United States for regulating GEOs and the products derived from them is the 1986 "Coordinated Framework for the Regulation of Biotechnology Products." This framework apportioned jurisdiction over transgenic products by using existing legislation and allowed the U.S. Environmental Protection Agency, the U.S. Department of Agriculture, and the U.S. Food and Drug Administration to work together in assessing the safety of the process and products of genetic engineering. In May 2000, the federal government conducted a six-month interagency review of its oversight of biotechnology products. The review explored the boundaries of the framework by focusing on several products that were not developed until the 1990s, and that therefore were not included in the 1986 framework. The review acknowledged that ensuring confinement could become one of the regulatory requirements for approval and commercialization of some GEOs. In 2001, the U.S. Department of Agriculture requested that the National Research Council's Board on Agriculture and Natural Resources (BANR) and Board on Life Sciences (BLS) review and evaluate bioconfinement of GEOs. BANR and BLS organized the Committee on Biological Confinement of Genetically Engineered Organisms.

The committee's charge was to review and evaluate bioconfinement methods and report on their application in confining transgenic crop plants, grasses, trees, fish, shellfish, and other organisms. The committee's report was to focus on genetic mechanisms, such as induced sterility, but it also was to identify and discuss other available or possible bioconfinement methods. The committee was asked to examine the following questions:

- What is the status of scientific understanding about various bioconfinement methods for genetically engineered organisms?
- What methods are available, and how feasible, effective, and costly are these methods?
- What do we know about when and why methods fail, and what can be done to mitigate those failures?
- When these methods are used in large-scale applications, what procedures can be used to detect and cull individuals for which the bioconfinement methods have failed? What is the cost effectiveness of these mitigation, detection and culling procedures?
- What are the probable ecological consequences of large-scale use of bioconfinement methods (e.g., deployment of sterile organisms) on wild populations, biological communities, and landscapes?
- What new data and knowledge are required for addressing any of these important questions?

Although not a focus of the report, the social acceptability of bioconfinement methods is discussed in the introduction and as context for the technical analyses.

This report examines bioconfinement of genetically engineered plants, animals, microbes, and fungi. Particular attention is given to transgenic fish and shellfish, trees and grasses, and microbes, because many of those species have been successfully engineered and currently are under federal regulatory evaluation. Because the committee was not asked to evaluate governmental practices or policy, it has limited its discussion to the scientific and societal components that are brought to bear on the process of choosing and applying bioconfinement of GEOs.

This report consists of six chapters. Chapter 1 defines terms and introduces concepts used throughout the report and briefly overviews the history and social acceptability of GEO confinement. Chapter 2 addresses the questions of when and why bioconfinement should be considered and it provides context for the need for and the application of bioconfinement methods. Chapters 3, 4, and 5 review and analyze bioconfinement methods for plants, animals, and microbes, respectively. Chapter 6 reviews the biological and operational opportunities and constraints for bioconfinement and examines bioconfinement failures and their mitigation. Chapter 6 concludes with a look to the future, exploring unanswered research questions that will establish better methods for the bioconfinement of GEOs.

RATIONALE FOR BIOCONFINEMENT

In many cases GEOs will not require bioconfinement, but in some cases they will. The need for bioconfinement should be evaluated on a case-by-case basis. The predominant factors for consideration involve the risks associated with the dispersal of a transgene or transgenic organism into a place, a population, or a biological community for which it was not intended. Significant research efforts on new categories of transgenic plants, insects, microbes, and animals are under way and many of those organisms are being considered for use or release into the environment. Species that disperse easily can pose particular risks because of the inefficacy of physical confinement methods and because of the potential for escapees to interact with and harm wild populations.

Currently, the most publicized environmental risk associated with transgene dispersal involves the evolution of increased weediness or invasiveness as a result of the sexual transfer of plant crop alleles to wild relatives. When domesticates and wild relatives live in proximity, it is not unusual for natural hybridization to occur. Spontaneous hybridization between nontransgenic crops and their wild relatives already has led to the evolution of several weeds and invasive species such as weed beets in Europe and weed rye in California. It is possible that some engineered genes that confer pest resistance or otherwise improve a crop plant might contribute to the evolution of increased weediness in wild relatives—especially if the genes escape to an organism that already is considered a weed.

A transgenic organism itself can become an environmental problem if the transgenic traits it expresses alters its ecological performance such that it becomes an invasive or nuisance species. Many crop plants pose little hazard because the traits that make them useful to humans also reduce their ability to establish feral populations in agricultural or nonagricultural habitats. However, feral and naturalized populations are well known for some crops and domesticated animals. If transgenes confer the ability to overcome factors that limit wild populations, the resultant genotype might be significantly more weedy or invasive than is its nontransgenic progenitor.

Other concerns about transgenic organisms include their effects on nontarget populations—including humans—and the potential for transgenes to disperse and spread before becoming deregulated in particular regions or nations. Gene flow from GEOs can greatly increase the extent to which nontarget species are exposed to novel proteins. In food crops, transgenes that code for "novel" pharmaceutical or industrial compounds could be candidates for bioconfinement if dispersal from their site of production is possible.

METHODS OF BIOCONFINEMENT

Many bioconfinement methods have been proposed for limiting the effects of transgenes. Although those approaches necessarily are tailored to specific organisms, and the terminology used to describe them is varied, all bioconfinement methods can be conceptually divided into three general categories: those that reduce the spread or persistence of GEOs; those that reduce unintended gene flow from GEOs into other organisms; and those that limit expression of transgenes.

Plants

Several existing methods target sexual and vegetative reproduction in plants. For example, sexual reproduction of genetically engineered plants can be blocked by including a gene that renders the organism either permanently sterile (nonreversible transgenic sterility) or sterile until the application of an appropriate trigger is available, such as the use of a chemical spray on a plant (reversible transgenic sterility). Other methods target pollen to confine pollen-mediated gene flow. Those include methods that achieve male sterility (the inability of a plant to produce fertile pollen) and those that transform chloroplast DNA—which usually is inherited maternally—rather than nuclear DNA. An alternative approach reduces the effects caused by unwanted transgenes by activating a transgenic trait through a specific artificial stimulus, such as a chemical spray (trait-genetic use restriction technology). A few of these and other methods are based on existing agronomic or horticultural practices and have been tested already; many others are newly developed or are only working hypotheses. Although the efficacy of some of the approaches is known, most are untested.

Animals

Confinement approaches in aquatic species can be achieved through physical confinement, through methods that prevent or disrupt sexual reproduction, or through methods that prevent GEO survival in the wild. The induction of triploidy is more established as a technique for finfish and mollusks than for crustaceans. Triploidy is a method that creates, in an organism, the state of having three sets of chromosomes in each cell nucleus, rather than the two typically found in most animal cells, which prevents successful cell division and reproduction. Triploidization is fairly successful and inexpensive, but like all bioconfinement techniques, it cannot guarantee 100% sterility. If only one sex of the GEO is used in the production operation—usually the female—then the likelihood that a self-sustaining feral population will become established is further reduced. All-female lines often are used for certain commercial species, and their use in conjunction with sterility techniques offers great promise. The use of single-sex lines is not a confinement system on its own, however, if related species that could mate with the GEOs are found nearby. If GEOs are crossed with related species, possibly sterile, interspecific hybrids would result, although thorough testing is required to ensure that sterility is close to 100%. Finally, several approaches could reduce the survivorship of GEOs by making them dependent on humans, either by genetically engineering the organism so that it requires an anthropogenic substance for its survival or by genetically engineering the organism so that it cannot live without an anthropogenic compound that blocks expression of the harmful gene.

There has not been much research on the bioconfinement of insect species, and so the subject is not well understood. Sterility is relatively easy to produce in insects by radiation, and the techniques used to produce and then quantify sterility are well characterized. However, sterility induced by radiation also can reduce the fitness of individuals. Any significant reduction of fitness would likely render the bioconfinement strategy ineffective within the target population, so transgenic approaches could be used to ensure sterility in insects without the loss of fitness. The large number of insects in any population, however, could make even a small failure rate of sterility techniques problematic.

Microbes

The two major bioconfinement methods used in microbes are phenotypic handicapping and the induction of suicide genes. The energetic cost of expressing the genetically engineered trait after phenotypic handicapping causes a loss in those organisms' ability to compete well with indigenous bacteria in soil and aquatic environments. Microbes multiply rapidly and can mutate, however, so subsequent generations of these bacteria might be better adapted to the environment than the original GE strain and be able to coexist with the indigenous populations. The ability to coexist depends in part on the highly variable external environment. Handicapped fungi appear to act similarly to handicapped bacteria in that they cannot compete with indigenous populations. The effectiveness of this mechanism as a confinement measure is unclear.

Another major confinement method for bacteria and fungi is the use of suicide genes. In the case of bacteria, numerous systems have been devised to significantly reduce the population of bacteria upon the addition or removal of a chemical or with a change in the environment. In the case of fungi, several mechanisms have been devised but none have been tested— even in the laboratory. In no case has any method been field tested. Suicide gene systems have not been used for viruses. Collectively, the methods remain completely theoretical or have been used in the laboratory, in small test plots, or in laboratory microcosms.

ENSURING BIOCONFINEMENT EFFICACY

Typically, precommercial evaluation of GEOs starts on a small scale and then is expanded to larger scales before release. It is likely that appropriate bioconfinement methods will be evaluated in a similar way. Nevertheless, these methods will vary in efficacy, depending on circumstances. Each is likely to work well with a specific organism, genotype, or environment, to work poorly with others, and to be inappropriate in yet others. Each is expected to work best on a small spatial scale, and the probability of failure increases with the number of individuals involved and the size of the area they occupy. Likewise, each is expected to work best over short periods of time, and the probability of failure increases with the amount of the time that the organisms are in the open environment. It is likely that no single method can achieve complete confinement on its own.

The efficacy of a bioconfinement method will vary depending on the organism, the environment, and the temporal and spatial scales over which it is introduced.

Most of the bioconfinement methods for GEOs discussed in this report are in the early stages of development and much is yet to be understood. It is clear that there is a great need for additional information on how well bioconfinement methods work separately and together. The effectiveness of any method will vary with genotype and the environment. Thus, the efficacy of combining confinement methods should be tested in representative genotypes that are under development and in environments into which a GEO is likely to enter to ensure that the plan is effective.

Before field release, the reproductive biology of the novel genotype should be measured relative to its progenitor to evaluate changes that might affect its rate of gamete and progeny production and dispersal. New genotypes generally do not have reproductive phenotypes that are different from those of their parents, but any changes that do occur could have important consequences. Changes in the reproductive biology of a GEO might not be anticipated because of its transgenes, but unanticipated phenotypic effects of a single allele are not rare.

• To evaluate changes in reproductive biology, the novel genotype should be compared with its progenitor before field release. For long-lived species, such as trees, it may be necessary to begin field tests before such comparisons are possible, with a realistic plan to mitigate any unexpected and dramatic increase in reproduction.

• Confinement techniques should be experimentally tested, separately and in combination, in a variety of appropriate environments and in representative genotypes under development before their application. In the case of long-lived organisms such as trees, they should be tested in conjunction with the release or scale-up of GEO products that are considered safe.

The evaluation of whether and how to confine cannot be an afterthought in the development of a transgenic organism. Safety must be a primary goal from the start of any project. Furthermore, it is important to consider the dispersal biology and the opportunities for the unintentional movement of transgenes in determining the best choice of an organism for use in creating a GEO. The constant and iterative evaluation of confinement options during the development of a GEO should optimize both the efficacy and the cost-effectiveness of the confinement options once they are deployed. Hurried consideration of confinement just before the deployment of a GEO will create a makeshift and expensive plan that might work better in theory than in practice.

The need for bioconfinement should be considered early in the development of a GEO or its products.

Many opportunities to mitigate the effects of a bioconfinement failure can be put in place during the earliest stages of development. For example, the act of choosing which GEOs to develop is in fact one form of bioconfinement. An organism that is typically grown to produce a common and widespread food product probably would be a poor choice as a precursor for an industrial compound unless that organism were to be grown under stringent conditions of confinement. This is an important issue for any novel compound or GEO for which zero tolerance of bioconfinement failure is needed. Engineering organisms that are not otherwise used for food or feed could be an effective way to prevent a transgenic compound from entering the human food chain.

Alternative nonfood host organisms should be sought for genes that code for transgenic products that need to be kept out of the food supply.

Given that no single bioconfinement technique is likely to be completely effective, the use of multiple techniques with different strengths and weaknesses will decrease the probability of failure. Many bioconfinement techniques are in the early stages of development, and some will be unacceptable in various circumstances. Therefore, before a GEO is released, its bioconfinement techniques should be tested in appropriate environments and in representative genotypes under development, and the reproductive biology of the GEO should be elucidated relative to its progenitor.

If a bioconfinement method is applied, the committee proposes that a new approach—an integrated confinement system (ICS)—should be used. ICS is a systematic approach to the design, development, execution, and monitoring of the confinement of a specific GEO. Among its features are a commitment to confinement by top management; the establishment of a written plan for confinement measures—and their documentation and remediation (in case of failed confinement); training of employees; assignment of permanent employees to maintain the continuity of the system; development and implementation of standard operating procedures; use of good management practices; periodic audits by an independent entity to ensure that practices are in place; periodic review adjustment to permit adaptive management of the system; and reporting to an appropriate regulatory body. For ICS to be effective, it is essential that it is supported by a rigorous and comprehensive regulatory regime that is empowered with inspection and enforcement.

An integrated confinement system that is based on risk assessment (including the risk of human error) is recommended.

There also is a need to define—early on—what constitutes adequate bioconfinement. This requires an evaluation of failures, their effects, and their probabilities under worst-case scenarios. It also entails the assumption that escaped genes have the opportunity to multiply.

The stringency of the integrated confinement system, including bioconfinement, should reflect the predicted risk and severity of consequences of GEO escape.

Because methods can fail, a single confinement method will not necessarily prevent transgene escape. For most GEOs, their escape will not pose a risk. In some cases, however, stringent confinement could be warranted, which a single method would not provide. Redundancy involves applying two or more types of safety measures to product design and use, each with fundamentally different strengths and possible vulnerabilities, so that the failure of one safety measure would be countered by the integrity of another. The choice of redundant confinement techniques, including bioconfinement, should consider a list of methods whose characteristics will combine to produce the best results. In many cases, this will involve the application of a mix of biological, physical, and physicochemical confinement measures tailored to specific GEOs. In other cases, it may be possible to combine two barriers of the same type but whose failures would be independent events, such that a failure of one barrier does not trigger a failure of the other.

- **It is unlikely that 100% confinement will be achieved by a single method.**
- **Redundancy in confinement methods decreases the probability of failing to attain the desired confinement level.**

The development, testing, and use of GEOs is increasing worldwide. GEOs can move across national borders by a variety of mechanisms including natural phenomena and trade. No country can manage all of the confinement issues that could affect its environment. An assessment of bioconfinement in any country will require attention to the efficacy of a given method and to concerns about its likely consequences—not just within that country but in other places as well.

- **Regulators should consider the potential effects that a failure of GEO confinement could have on other nations, as well as how foreign confinement failures could affect the United States.**
- **International cooperation should be pursued to adequately manage confinement of GEOs.**

A bioconfinement scheme will be effective only if it is fully implemented, and several factors affect compliance. The efficacy of bioconfinement will vary with the human processes involved in applying the technique; the confinement method itself; the characteristics of the GEO; the cost of compliance; the characteristics of the organizations involved; the regulatory system in place; and public transparency.

The majority of the bioconfinement methods discussed in this report are in development and have not been used in conjunction with commercially available GEOs. Consequently, the public has had little opportunity to develop opinions regarding this aspect of biotechnology. Nonetheless, GEOs or their products can have social significance or be infused with symbolic, social, and aesthetic values that might present important challenges for determining the need for or application of bioconfinement methods. In order to enhance public trust and acceptance of a given confinement strategy for a GEO, a sound science-based risk assessment might need to be coupled with a clear and public articulation of any potential ethical concerns.

Broad social and ethical values should be considered in assessing the stringency of the integrated confinement system which includes bioconfinement.

The public's right to information—often called *transparency*—and its right to participate in decision making are fundamental to the practice of democracy. Each right complements the other. Appropriate transparency and public participation can improve the effectiveness of confinement, for example, by informing decision makers about otherwise unknown facts about the environments in which confinement would be implemented, and can increase the acceptance of bioconfinement measures (and of the GEOs being confined) by building trust in the decision-making process.

Transparency and public participation should be important components in developing and implementing the most appropriate bioconfinement techniques and approaches.

DETECTING AND MITIGATING BIOCONFINEMENT FAILURE

Failures in the bioconfinement of GEOs have not been documented to date, in part because so few methods have been implemented. However, given the imperfections of methods under development and those of methods that have been applied to nonengineered species, it is likely that failure will occur. The degree to which failed confinement events can be monitored and managed depends on whether the GEOs are easily detected, the scale at which they are released into the environment, the GEOs' subsequent population dynamics, and the degree to which they can hybridize with related species. Early detection of failed methods will be important for mitigating bioconfinement failure, especially if the confined transgenes are likely to spread. Even if a failure is detected early, effective mitigation might not be feasible.

Some limited options are available for detecting individuals and culling them after failed bioconfinement. In plants, a failure might be signaled by a distinctive phenotypic trait, such as the presence of flowers on plants that have been engineered to lack them, so workers could cull abnormal plants from small fields. The failure of many bioconfinement methods, however, will be much more difficult to detect. For example, elaborate experiments would be needed to determine whether a repressible seed-lethal transgene is functioning properly. Also, many bioconfined plants will be grown on such large areas of land that repeated comprehensive inspections will be impractical. In the future, DNA "fingerprints" could be linked to bioconfined transgenes to function as "bio-barcodes" that could be detected and used to cull GEOs. Remote sensing approaches might also be available to detect GEOs.

It is feasible to detect and then cull individual fish in which triploid sterilization induction fails before they are transferred from secure hatcheries to much less secure facilities, such as outdoor ponds or open-water cages. Economies of scale and possible automation could reduce the cost of such efforts. A similar approach can be applied to oysters and shrimp. To detect and cull failures in bioconfinement of fish, shellfish, or insects, one could screen for proteins expressed by the key gene involved or for a co-inserted marker gene. Nonlethal detection might be possible for larger organisms or with such marker genes as green fluorescent protein; detection in smaller organisms—especially insects—would be more likely to require lethal sampling. It is not currently possible to detect or cull microbes if bioconfinement fails. The committee did not speculate about cost-effectiveness because genetic engineering-based bioconfinement methods are theoretical or at an early stage of development.

Current methods for detecting and culling individual GEOs after a bioconfinement failure are very limited, and they depend on the organism and scale of the original release of the GEO.

For large-scale GEO releases, effective monitoring will be essential for mitigating failure. Currently, monitoring is difficult because it involves searching for what often will be a rare event over a potentially large area. In the future, organisms might be purposefully transformed with additional constructs for monitoring. Ideally, monitoring methods would be developed that could identify escapes with remote sensing. Monitoring should be seen as a complement to confinement—not a replacement for it. That is, the act of monitoring should not result in complacency about the possibility of a bioconfinement failure.

Easily identifiable markers, sampling strategies, and methods should be developed to facilitate monitoring of bioconfined GEOs in the environment.

ECOLOGICAL CONSEQUENCES OF LARGE-SCALE USE OF BIOCONFINEMENT

Many bioconfinement methods might be successfully used and result in certain GEOs having negligible effects on wild populations, biological communities, or ecosystems, but there has been little research on this topic. Some methods have been used in nonengineered organisms in the past, often with other goals, and they were considered in the committee's evaluation. Those methods include growing male-sterile crops for hybrid seed production, small-scale rearing of sterile fish, and releasing sterile male insects that mate with wild females as part of biocontrol strategies to reduce pest insect populations.

Two related areas of ecological concern about the use of bioconfinement with GEOs were identified: the large scale at which bioconfined organisms could be released and the possibility that even carefully planned, integrated bioconfinement methods could fail. In some cases, the area over which sterile or handicapped GEOs are released could be large enough to affect biodiversity. In salmon and other species, the presence of large numbers of sterile GEOs or those with reduced fitness in some cases could threaten local biodiversity. There is concern that some native populations of animals might lose the ability to compete for food or mate successfully in the presence of more competitive or more attractive, but sterile, GEOs. If this were to occur in small populations, depressed levels of natural reproduction could threaten the long-term survival of native genotypes. In other cases, the large-scale release of sterile GEOs could have the beneficial effect of alleviating existing problems, such as the loss of genetic diversity that can occur when modern—and often genetically uniform—crop plants (or hatchery-raised fish) interbreed with rare wild relatives or locally adapted varieties.

A more general problem with all bioconfinement methods is that occasionally they could break down, especially if they are intended to confine millions of free-living individuals. Depending on the original reasons for using bioconfinement, the ecological consequences could be serious. If a bioconfined GEO can become a pathogen or an invasive species after the breakdown of an Integrated Confinement System, the decision to release it on a large scale should be scrutinized with extreme caution. If the reason for using bioconfinement is mainly commercial, the ecological effects of bioconfinement failure could be of no consequence. It is difficult to generalize about the ecological effects of large-scale releases of bioconfined GEOs, and further research should address these questions in relation to specific realistic conditions.

Research is needed to characterize potential ecological consequences of bioconfinement methods and to develop methods and protocols for assessing environmental effects should confinement fail.

CONCLUSIONS

The current lack of quality data and science is the single most significant factor limiting our ability to assess effective bioconfinement methods. In many cases GEOs will not require bioconfinement, but when they do the need for bioconfinement should be evaluated case by case, considering worst-case scenarios and the probability of their occurrence. The evaluation of whether and how to confine a GEO should be an integral part of its development, and the need for bioconfinement should be considered early in the process. It is unlikely that any single bioconfinement technique will be completely effective, and using multiple techniques with different strengths and weaknesses will decrease the probability of failure. Furthermore, many bioconfinement techniques still are in the early stages of development, and the possible unintended consequences of some bioconfinement methods mean that some technologies will be unacceptable under certain circumstances. Therefore, before a GEO is released, the techniques to be used should be tested in a variety of appropriate environments and in representative genotypes under development, and the reproductive biology of the GEO should be understood relative to that of its progenitor. If a bioconfinement method is applied, an integrated confinement system should be put in place. Such a system must be supported by a rigorous and comprehensive regulatory regime empowered with inspection and enforcement.

Finally, in order to implement effective bioconfinement of GEOs, the committee recommends support for additional scientific research that

- characterizes as completely as possible the potential ecological risks and consequences of a failure in bioconfinement
- develops reliable, safe, and environmentally sound bioconfinement methods, especially for GEOs used in pharmaceutical production
- designs methods for accurate assessment of the efficacy of bioconfinement
- integrates the economic, legal, ethical, and social factors that might influence the application and regulation of specific techniques
- models (using models that are calibrated and can be verified experimentally) the dispersal biology of organisms targeted for genetic engineering and release, where sufficient information does not exist.

Interdisciplinary research will improve the future of biotechnology by developing new confinement methods that minimize the potential for unintended damage to human health and the environment. The success of these efforts will do much to bolster public confidence in the continued growth, development, and opportunities presented by biotechnology.

1

Introduction

WHAT ARE GENETICALLY ENGINEERED ORGANISMS?

The human community has engaged in genetic modification for thousands of years with the domestication and subsequent breeding of microbes, plants, and animals for use in agriculture, medicine, and industry. The developments of the past 25 years, which involve the insertion and manipulation of genes within an organism's DNA, however, constitute a significant advance from the process of selective breeding. Four major parameter shifts from selective breeding to genetic engineering illustrate both the power of the new methods and the controversy that surrounds their application (Kreuzer and Massey, 2001). First, selective breeding operates on the whole organism, so factors, such as generation time and development patterns, determine the speed with which a trait or characteristic is selected. Second, selective breeding is less precise, inevitably moving sets of genes that are linked to those targeted for introgression. Many of those linked genes have unknown functions. Third, the certainty surrounding genetic expression is low for selective breeding; genetic change is often poorly characterized. Finally, the genes and traits that can be used for genetic improvement typically come only from the same species or from one that is closely related (Kreuzer and Massey, 2001).

In contrast, genetic engineering operates at the cellular or molecular level so it is possible to select and transfer single genes. The genes of interest generally are well characterized, and they can come from other species, including those from distant taxa.

Several definitions of genetically engineered organism (GEO) have

14

emerged over the past few years from an assortment of institutions and policymakers, but there appears to be a general understanding that a GEO is "an organism that has been modified by the application of recombinant DNA technology" (FAO, 2002). Although GEO and GMO (genetically modified organism) often are considered interchangeable terms, this report uses GEO to be consistent with recent reports of the National Research Council (NRC, 2002a; 2002b).

WHAT IS BIOCONFINEMENT?

Since the first release of GEOs into the environment in the mid-1980s, public and scientific concern has focused on the potential consequences of the escape of those organisms and their associated transgenes into natural and managed ecosystems. It has long been asserted that GEOs could not compete successfully with wild populations, and therefore that they could not survive in the wild over the long term. Recent and longer standing ecological studies might suggest otherwise, however—especially if GEOs can cross with wild relatives (Linder et al., 1998; Snow et al., 2003). As a result, there is interest in developing methods to confine some GEOs and their transgenes to designated release settings. There also are cases in which the movement of transgenes from one domesticated plant or animal variety to others must be confined. Many confinement methods are biological, and they are referred to as *bioconfinement* in this report.

Bioconfinement of GEOs is in the conceptual and experimental stages, although some methods have been applied to control nonengineered organisms. Bioconfinement includes the use of biological barriers, such as induced sterilization, that prevent GEOs or transgenes from surviving or reproducing in the natural environment (Chapter 3). More specifically, sterility has been induced in some species of salmon and oysters through manipulation of the number of chromosomes in individual animals (Chapter 4). It also is possible to introduce a single genotype of a self-incompatible plant to prevent seed formation. This practice is used in horticulture and agriculture (Chapter 3).

There is a long list of potential techniques and principles for the bioconfinement of GEOs (Chapters 3, 4, and 5). How important they become will depend on many factors, as this report outlines. The Committee on Biological Confinement of Genetically Engineered Organisms notes that this report has been prepared in the early days of those emerging techniques. Over the course of preparing this report, the committee informally surveyed several representatives from the private sector about emerging bioconfinement methods. While not a comprehensive survey, the committee came away with the impression that, at this time, industry research efforts on new bioconfinement methods were fairly modest and that current efforts mainly focused on utilizing and refining existing well-established methods of confinement (such as physical, spatial, and temporal isolation) with GEOs under development. The bioconfinement principles and methods discussed here are likely to be improved and augmented through research and development over the years to come.

OTHER CONFINEMENT METHODS

It is useful to consider different confinement measures because every type has an inherent vulnerability to failure. The two other broad types of confinement measures, the *physical* and *physicochemical barriers*, prevent the escape of organisms or their genetic material (via gene flow to relatives) from the production system into accessible ecosystems. A brief introduction appears below, and a more detailed discussion of issues to consider in applying those barriers to the confinement of genetically engineered organisms appears in a document by the Scientists' Working Group on Biosafety (1998).

Physical barriers are devices, such as screens, that prevent organisms at a given life stage (gamete, asexual propagule, juvenile, adult) from leaving the production operation or facility. For annual crops and for insects, physicochemical barriers include screens with appropriate mesh sizes on windows and other openings of greenhouses. Another effective barrier for confinement of plants and insects is the use of negative air pressure achieved when the volume of air exiting a space or chamber exceeds the air intake volume to contain pollen, spores, or mobile insects (Traynor et al., 2001).

Physical barriers for fish, shellfish, and algae produced in aquaculture systems include screens in pipes and channels of water that flow in and out of ponds or tanks, effluent drain structures with multiple mechanical barriers (French drains), and the fine sand filters that often make up one component of closed-loop aquaculture systems. It is particularly important to match the design and operation of a mechanical barrier to the smallest life stage it is expected to restrain, keeping in mind that gametes or asexual propagules of aquatic organisms can be miniscule (viable eggs and newly fertilized embryos with diameters less than 10 micrometers). Schematic diagrams of physicochemical barriers—along with examples of operating criteria designed to hold back specific life stages—appear in two scientific biosafety guides (Agricultural Biotechnology Research Advisory Committee, 1995; Scientists' Working Group on Biosafety, 1998).

Physicochemical barriers induce mortality through lethal physical alterations to the escape routes to the environment immediately external to the site of GEO production with the aim of achieving 100% mortality. Physical barriers applied in the production of genetically engineered fish, shellfish, and algae include the use of temperature changes, changes in pH, or the addition of dissolved chlorine to water that flows out of fish tanks or ponds before the effluent is discharged into the environment. Typically, the effluent water passes through a chamber that imposes the lethal condition for a given contact period and then, before the effluent is discharged to a natural water body, the effluent water is restored to ambient environmental conditions to maintain the water quality of the receiving water. It is fairly easy to impose lethal physical conditions because most farmed aquatic species have a well-known and narrow range of physical parameters needed for survival. The mechanisms for altering these are well understood from many years of industrial and municipal water treatment. Water temperature changes are easily achieved by heating and cooling and pH can be adjusted using acids and bases. Chlorine, even at low levels (one part per million), is lethal to most organisms.

Bioconfinement Redundancy

In many technology applications, the principle of "redundancy" guides efforts to reduce the probability that predictable hazards will occur, and thus achieve the benefits of technology. Generally, two or more safety measures are applied to product design and use, each with fundamentally different strengths and vulnerabilities, so that the failure of one will be balanced by the integrity of another. A unique feature of biotechnology that distinguishes it from other recent technologies is the fact that it involves living organisms and products. As such, biotechnology, like all biological systems, inherently operates with a given level of uncertainty. This attribute makes the application of the principle of redundancy particularly relevant to bioconfinement as well. In many cases, redundancy will involve the application of an appropriate mix of biological, physicochemical, and mechanical confinement (Agricultural Biotechnology Research Advisory Committee, 1995; Kapuscinski 2001; Scientists' Working Group on Biosafety, 1998). In other cases, it may be possible to combine two barriers of the same type but whose failures would be independent events, such that a failure of one barrier does not trigger a failure of the other.

One application of the principle of redundancy in aquaculture of genetically engineered (GE) fish combines physicochemical barriers (floating cages that are highly prone to failure and land-based, flow-through units that are less prone to failure) with biological confinement consisting of production of an all-female line of sterile fish.

SCOPE OF THE REPORT

This report reviews biological methods used to confine genetically engineered organisms. It focuses on the genetic mechanisms of bioconfinement, such as induced sterility, but it also identifies and discusses other available or possible methods. The following specific questions are addressed:

- What is the status of scientific understanding about various biological confinement methods for genetically engineered organisms?
- What methods are available, and how feasible, effective and costly are these methods? (e.g., How well would these methods fit with existing practices for research and agricultural production? When and for what systems are the individual methods appropriate?)
- What do we know about when and why methods fail, and what can be done to mitigate those failures?
- When these methods are used in large-scale applications, what procedures can be used to detect and cull individuals for which the biological confinement methods have failed? What is the cost-effectiveness of these mitigation, detection, and culling procedures?
- What are the probable ecological consequences of large-scale use of biological confinement methods (e.g., deployment of sterile organisms) on wild populations, biological communities, and landscapes?
- What new data and knowledge are required for addressing any of these important questions?

Although not a specific focus of the report, the social acceptability of bioconfinement methods is discussed in the introduction and as context for the technical analyses.

This report examines a variety of issues associated with bioconfinement of transgenic fish and shellfish, trees and grasses, insects, and microbes. Fish, such as transgenic farmed salmon, could pose special environmental risks because of the inadequacy of physical confinement methods (net pen enclosures) and because of the potential for escapees to interact with and harm wild fish stocks—many of which already are in decline. There is concern about the gene flow of trees and grasses related to their high pollen production and the presence of sexually compatible wild species. Those plants are perennials, and environmental exposure issues could differ from those associated with corn, soybean, or other annuals. Concerns regarding perennials are bolstered by the longevity of individual plants and by characteristics that inhibit growth of other species, including—in the case of trees—their large, shade-producing physical structures and the accumulation of surface litter they cause.

There have been field tests of transgenic insects and animals, and more are under consideration, but several genetically engineered crops are now in relatively common use. Algae, plants, mammals, insects, shellfish, and microbes are being genetically engineered in the laboratory and are now or could someday be considered for release into the environment. Although each species will have unique characteristics that determine the effectiveness of the bioconfinement methods applied, there are some general principles that could be applied to a common framework for safe use.

INTERNATIONAL ASPECTS

Although much of the discussion of confinement of GEOs occurs in a domestic context, several significant international dimensions are of interest. Development, production, and use of GEOs is on the rise in other nations—including Argentina, Brazil, China, Canada, Cuba, and India—in part because of the global character of the biotechnology business, which can transport research, field testing, and production from the United States to other nations as it becomes expedient to do so. Business enterprises could choose between regulatory or intellectual property regimes, for example, and move GEOs with confinement techniques from an environment in which they are suited to one in which they are not. Similarly, individual people or businesses desiring to use a GEO could import or export that organism for their own purposes. GEOs can be moved between countries by any number of means—international trade, travel, tourism, transport, and aid, for example—as well as unintentionally in ocean and river currents, wind, storms, and floods. Animals such as birds, insects, and rodents could be vectors, and the organisms themselves can move across borders.

As a matter of United States public policy, addressing bioconfinement thus requires that the issues of efficacy, public concern, and environmental consequences be considered as they pertain here and abroad. Similarly, the United States has an interest in bioconfinement policy—and practice—in countries from which GEOs could come and in the effectiveness of international regulation of such movement. No one nation can control all of the confinement issues that could affect its environment, in part because of the dispersal of GEOs across national boundaries. The committee's finding is that adequately addressing bioconfinement may require international cooperation.

HISTORY OF CONFINEMENT

The history of genetic engineering is coextensive with the history of confinement and containment methods. The first recombinant DNA molecule was engineered by researchers led by Paul Berg at Stanford in 1972. They isolated and employed a restriction enzyme to cut DNA from two different viruses—the bacterial virus, lambda, and the mammalian virus, SV40—and used the enzyme ligase to paste two DNA strands together to form a hybrid circular molecule. The goal was to use the hybrid virus as a vector to deliver genes to bacteria. In a short period of time, several recombinant DNA (rDNA) organisms were created and vectors were developed. The research raised fears about the danger posed by the new organisms. Berg suspended the work in 1972 in response to charges that the risks to laboratory workers and to the general public were unknown and potentially grave (Wade, 1979; Wright, 1994). The use of *Escherichia coli* (*E. coli*) as a model organism exacerbated those concerns.

The scientific community was divided on the potential dangers of research with GEOs, and there was no accord on the appropriate steps to take. Several committees were formed in the United States and in other countries, including one convened by the National Academy of Sciences (NAS) (Wright, 1994; for the letter to NAS that instigated NAS action, see Singer and Soll, 1973). At the same time, activists and nongovernmental organizations (NGOs) began a campaign of opposition that called on the government to regulate, control, and limit the action of scientists in this area. Most of the scientific community saw tremendous potential in genetic engineering, both for basic science and the practical developments to which it would lead, including therapeutic biological agents like human insulin, genetically engineered crops and animals, and methods of gene therapy. In response to concerns about the hazards of the research, a moratorium was imposed (with strong support from the National Institutes of Health [NIH]) on some experiments. In 1975, an international meeting in Asilomar, California led to consensus on the importance of developing safety measures for use in the genetic engineering of bacteria and viruses (Wade, 1979; Wright, 1994). The primary concerns at the time involved the possibility that genetically engineered bacteria could develop drug resistance or have other traits harmful to humans. There was concern that the accidental release of altered bacteria could have disastrous effects on public health. There was additional worry that as a result of their modification, harmful viruses could extend their host range to humans.

"Biological containment" was one of the first mechanisms proposed to control the risks of the new technology. Among the first groups to address the risks of recombinant DNA research in Great Britain was the Ashby Committee, which was organized by the Advisory Board for the Research Councils. In written testimony to that committee, Sydney Brenner of the Cambridge Laboratory for Molecular Biology warned of the dangers of having potentially dangerous materials handled by improperly trained scientists. In 1974 he proposed creating bacteria that were genetically engineered not to survive outside of the laboratory, to reduce the possibility of those organisms transferring their DNA to other organisms. The suggestion was endorsed in the Ashby Committee report. This was the first time the concept of biological confinement was introduced in an official report (Wright, 1994).

The attempts at self-regulation by the scientific community culminated in the 1975 Asilomar meeting and in the creation of the NIH Recombinant DNA Advisory Committee (RAC). Soon after the Asilomar meeting, the first RAC meeting took place to construct systematic safety guidelines. Agreement proved difficult, and it took 16 months (until June 1976) for consensus on safety protocols for assessing the degree to which a particular genetically engineered organism represented a hazard and then applying two independent confinement systems—one physical, the other biological. The physical confinement systems were defined in terms of standard techniques of microbiology research that were in common use in U.S. laboratories. Laboratories were identified on a four-point scale: from BL1-BL4. BL1 facilities were standard microbiology laboratories that used no special safety procedures. A BL2 rating represented little more. Aerosols were to be confined to cabinets and eating and drinking were prohibited. A BL3 rating required the use of special procedures, equipment, and design at an estimated cost in 1977 of $50,000 for a typical facility (Wade, 1979). Included in the BL3 designation were cabinets which controlled of air flow out of, but not into the laboratory. BL4 laboratories were those in which extremely pathogenic agents, such as smallpox or Lassa fever virus, were studied. Costing about $200,000, a BL4 laboratory was less expensive to build from the ground up than to create through modification of an existing facility.

In addition to requiring physical confinement, the RAC protocols listed the first systematic assessment of biological confinement, graded on a scale from EK1 to EK3. For EK1, researchers would use the standard strain of *E. coli* (K12) as a host. That strain is not very robust after multiple generations of laboratory rearing. The only vectors authorized for use in EK1 were plasmids that had a low probability of transferring DNA to other bacteria. The suggestion of the Ashby Committee was taken up for EK2 conditions. In an EK2 laboratory, the genetically engineered organism and the vectors would be versions of the handicapped bacteria further engineered to have only a 1 in 100 million chance of survival outside the laboratory, according to laboratory tests. For example, researchers developed a strain of *E. coli* that could survive only in the laboratory because its survival depended on specific chemicals that do not occur commonly elsewhere (Curtiss, 1978, see Box 5-1). EK3 laboratories required genetically engineered hosts for which the laboratory findings of EK2 organisms had been confirmed through actual feeding to animals, humans, or both.

The system of safeguards came under criticism both from within and outside of the scientific community. Several prominent scientists, and many nonscientists, charged that the entire discussion and protection system focused far too narrowly on health risks, and that it failed to address broader ethical and evolutionary issues. Many scientists pointed out that the physical confinement system depended too much on the behavior of the people working in laboratories.

Battles over the guidelines and whether legislation would be needed—particularly with regard to non-NIH-funded recombinant DNA research in the private sector continued for several years. As a result the RAC evolved to include members of the general public. Eventually, the increase in data on the safety of handicapped genetically engineered model organisms and the growing recognition of the benefits of the emerging technology, both commercial and therapeutic, quieted the debate. Many of the requirements for bioconfinement were relaxed, and research was allowed to proceed. The guidelines were broadened to include any institution receiving NIH funding (regardless of the source of funds for a given experiment) and some commercial laboratories, which were subject to limits imposed by considerations of liability (Uchtmann, 2002). The philosophy during this period relied more on establishing professional norms and voluntary compliance than on extensive government regulation as the basis of protecting the public (Uchtmann, 2002; Wright, 1994).

In the 1980s, GEOs moved from the laboratory to the field, requiring a major shift in confinement strategy. The NIH guidelines initially prohibited "deliberate release." The first attempt to obtain RAC approval for a field trial involved "ice-minus" bacteria. Researchers at the University of California, Berkeley, had created a strain of *Pseudomonas syringae* that could reduce frost formation on plants where it replaced naturally occurring populations of bacteria (Sprang and Lindow, 1981). The first request for field trials in 1982 was postponed, but approval came in 1983. Activists filed a lawsuit and successfully halted the trials. Once more, genetic engineering produced a rancorous debate. The media, the courts, and the scientific community voiced their opinions about the risks and benefits of general release of GEOs into the environment and about how the regulation and local laws should work (Uchtmann and Nelson, 2000). The legal maneuverings and political debates delayed field trials for another four years. By 1988, although the trials showed that the bacterium could reduce ice formation, plans to commercialize the product were abandoned.

Clearly, a coordinated regulatory system was needed for field testing GEOs. The 1984 National Environmental Policy Act (NEPA) required environmental assessment of the impact of any action taken by RAC—and hence by NIH—that might affect the environment. By 1986, a coordinated framework, built on existing laws and institutions, had been developed to satisfy NEPA (51 Fed Reg 23302 June 26, 1986 described further in Chapter 2). Henceforth, regulation would focus on the products of genetic engineering rather than on the process. The degree of confinement necessary for approval under NEPA would depend on the traits that were to be introduced into an organism and on the results of an environmental assessment. For many types of crop plants, such as herbicide-tolerant soybean, no confinement would be required by US regulatory agencies.

By the early 1990s, researchers were developing some of the biological confinement methods described later in this report. The difficulty of relying on physicochemical confinement methods for fish, in particular, led to the development of new methods, including sterile triploidy, antifertility genes, and "suicide genes" (Aleström et al., 1992; Donaldson et al., 1993). International discussion about transgenic organisms and the confinement methods that would be required to ensure the environmental safety and health of various populations proceeded along the lines of the initial Asilomar-based framework. In Oslo, Norway, the First International Symposium on Sustainable Fish Farming similarly led to discussion of the need for international agreement on the use of bioconfinement to supplement physical barriers (Aleström, 1995).

An important development in the technology of bioconfinement was the invention in the late 1990s of the Technology Protection System (Chapters 3 and 6) by Melvin Oliver—a U.S. Department of Agriculture (USDA) researcher—who created it in conjunction with Delta and Pine Land Company. Its purpose was to protect the intellectual property rights of biotechnology companies that develop seeds for crops. In partnership with Monsanto, the Delta and Pine Land Company was working to develop a strain of cotton that, with the application of the Technology Protection System (TPS), would not produce usable seed at the end of each growing season. The companies saw the potential to apply the technology to a variety of genetically engineered products, making it a potentially valuable commodity in itself. Delta and Pine Land (and later Monsanto) anticipated potential use of the technology even for corn.

As genetically engineered food crops came into common use in the United States throughout the 1990s, with no stipulations about confinement, several new concerns emerged about cases for which confinement would be desirable. First, there was concern that those crops would introduce transgenes into plant and animal populations for which there was a strong social, ethical, or economic interest in maintaining non-GEOs. For example, farmers of organic products were worried that pollination from genetically engineered plants would introduce transgenes into their crops and threaten their ability to sell their harvest as certifiably organic. Reports that transgenes were found in indigenous landraces of maize in Mexico (Alvarez Morales, 2002; Quist and Chapela, 2001) fueled several conflicting controversies (e.g., Christou, 2002; Martinez-Soriano et al., 2002; Metz and Fütterer, 2002; Kaplinsky et al., 2002; Quist and Chapela, 2002), including the question of how deregulation of a crop in the United States is viewed by other nations.

The possibility that crop genes or crop products not approved for human consumption could enter the food supply brought attention to other problems with existing confinement methods, such as spatially separated fields. The difficulty of segregating commodity crops was made apparent in the case of StarLink corn (see Box 2-1). StarLink produces the *Bacillus thuringiensis* toxin protein Cry9c. In 1998, the United States Environmental Protection Agency (US EPA) approved StarLink corn for animal consumption and for industrial production of ethanol. Human consumption was not approved because Cry9c resists both heat and digestion, and those traits are associated with allergens. Thus, US EPA determined that the protein was itself a potential allergen. In September 2000, several newspapers reported that StarLink corn had been detected in Taco Bell brand taco shells sold in grocery stores. The Genetically Engineered Food Alert, a coalition of environmental groups, had sent the shells to the Iowa-based company Genetic ID for testing. That independent laboratory reported that the sample taco shells contained at least 1% StarLink corn. Kraft Foods, which distributed the taco shells, responded the next day with a press conference and a "special report" posted on its web site. Kraft stated that it was conducting its own tests to confirm the results, and would voluntarily recall the taco shells if Cry9C was detected.

Kraft later confirmed that Cry9C was in the taco shells, and it recalled the nearly 3 million boxes. Subsequently, hundreds of corn-based products were recalled because of concern about StarLink contamination. In late September 2000, Aventis—the company that developed and produced the seed—suspended sales. That was the first time a biotechnology company had frozen sales of a genetically engineered seed. By agreement with USDA, Aventis bought back all the remaining StarLink corn to ensure that it would be used only for animal feed and ethanol production (at a cost of roughly $100 million). There is no evidence that StarLink produced an allergic reaction in any person.

Newer applications of GEOs include plant-made pharmaceuticals (PMPs) and crops that are used in production of industrial compounds. Given the large start-up costs of factories that produce biologics, such as human monoclonal antibodies, PMPs represent a potentially lucrative market. But the developments also heighten worries about the escape of transgenes or transgenic plant products into the food supply. Among the first companies to come to public attention was Prodigene, a leader in PMP development. In 2001, Prodigene planted corn genetically engineered to produce pharmaceutical products at various field sites. The next year, at one of the Nebraska sites, a conventional crop—soybeans—was grown on the land that had been used for the experimental crop. Seed from the experimental crop germinated among the soybean plants. Although volunteer corn initially went undetected, eventually, inspectors from the USDA Animal and Plant Health Inspection Service identified the plants. Nonetheless, the plants were subsequently harvested with the soybeans. Pieces of transgenic corn plants ended up with the soybeans in a grain elevator. USDA imposed a quarantine of 500,000 bushels of soy. Some commentators hold that this demonstrates success of the current regulatory system; others argue that it shows that companies cannot be trusted to apply adequate protection. In either case, the importance of confinement methods has been demonstrated as have their potential weaknesses (Taylor and Tick, 2003).

SOCIAL ACCEPTABILITY OF BIOCONFINEMENT METHODS

While the need for effective confinement methods for some types of GEOs has become more apparent in recent years, the majority of the bioconfinement methods discussed in this report are in development and have not been used in conjunction with commercially available GEOs. Based on the committee's best judgment and collective expertise, it appears that the public has had little opportunity to develop opinions regarding this facet of biotechnology. It is likely that the public's acceptance of GEOs and bioconfinement will be closely linked or correlated, and that some GEOs could receive greater acceptance based on the confinement method associated with them. In other cases, GEOs could be viewed as less acceptable or even potentially dangerous because the associated confinement method indicates the serious risk posed by the GEO. It is premature to make predictions regarding public acceptance. However, the following case study of one bioconfinement method could provide some insight into the public's future response to bioconfinement.

Case Study of the Technology Protection System: "Terminator"

At the time the Technology Protection System was being developed, some seed manufacturers were requiring their customers to sign contracts prohibiting them from saving and reusing seed from cultivars with utility patents. Compliance was an issue, and the Technology Protection System seemed an ideal solution. The TPS was created through a Cooperative Research and Development Agreement (CRADA) signed between the U.S. Department of Agriculture's Agricultural Research Service and the Delta and Pine Land Company in 1993, and the developers jointly received a patent on the process in 1998 (U.S. Patent 5,723,765).

The Rural Advancement Foundation International (RAFI; now the Action Group on Erosion, Technology, and Concentration), an NGO based in Canada that opposes the use of biotechnology, labeled the Technology Protection System "terminator technology." The NGO drew considerable attention to a potential impact of the system if it were to be applied: subsistence farmers who traditionally saved seed from one season to the next no longer would have that option. Seeds collected from their technology-protected GE crops would not be viable.

The traditional farming practice of seed saving is widespread and important to farmers throughout the world. The loss of the ability to save seed—which could compromise food security for resource-poor farmers—was a concern with terminator technology. This concern resonated with the mainstream international media which helped publicize the issue. The concept of terminator technology and concern about its consequences for subsistence farming helped make this story much simpler to report than were other stories about genetic engineering (Lambrecht, 2001; Priest, 2001).

The issue of seed saving had other ethical implications. A seed can be viewed as a living organism—and reproduction is an essential part of what living organisms do (Boorse, 1975). When one purchases a product, it is usually implicit that one is entitled to the full, normal functioning of that product (Cummins and Perlman, 2002). If reproduction is understood as part of the normal functioning of an organism, then that claim would extend to the offspring of the organism. Terminator technology for intellectual property protection has faced opposition for this reason (Eaton et al., 2001; Halweil, 2000).

Widespread public debate about terminator technology ensued, and objections came from the Consultative Group on International Agricultural Research, whose members unanimously recommended banning research on terminator genes. Further criticism came from Gordon Conway, president of the Rockefeller Foundation (Lambrecht, 2001). Monsanto (who had sought to acquire Delta and Pine Land in part because of its 1998 patent on the system) subsequently announced that it dropped the technology.

The potential impact on subsistence farmers was not the only concern associated with the terminator technology. The campaign of RAFI also claimed that the development of terminator technology was evidence that "life-science companies were bent on controlling the food chain" (Lambrecht, 2001). This argument is part of a larger, continuing debate about whether "biotechnology will help a few well-capitalized companies control decision making in agriculture and limit farmers' ability to choose from an array of production possibilities" (Thompson, 2000). Terminator and other "genetic use restriction technologies," also known as GURTs (see Chapter 3), have been linked with corporate interests in protecting intellectual property rights. The continued legal battles involving infringement reinforce this association, even in cases that do not involve such technology (Priest, 2001). Although there were sporadic attempts to highlight the utility of terminator technology as a tool for bioconfinement, there is no evidence that this was taken seriously, either as a motivation for the initial development of the technology, or by environmentalists. More recently, similar technologies have been developed for use in confinement rather than strictly for protection of intellectual property rights (McHughen, 2000).

In summary, a lesson from the terminator technology experience is that social acceptability based on ethics can be a powerful influence on the decision to adopt or reject a bioconfinement method. A combination of efforts by numerous NGOs, activist organizations, and the media generated enough public opposition to terminator technology to persuade several companies and agencies to abandon it. In the future, will the response to related technologies and other reproductive methods of bioconfinement be different from the rejection of terminator technology? Are there alternative approaches to developing and characterizing bioconfinement methods that could be met with greater acceptance than terminator technology? This report describes many possible options for bioconfinement—with the goal of stimulating constructive debate and discussion.

Consequentialism and Public Acceptance

The potential risks associated with the bioconfinement of GEOs can be understood within the framework of a major philosophical tradition—consequentialism—which defines what is right for society in terms of maximizing the net good. For confinement methods, acceptability is determined with objective values used in risk-benefit analyses. The consequentialist approach is implicit in much of the United States regulatory system, and is the reason for the focus on scientific risk assessment to analyze uncertainties with GEOs. However, as the experience with terminator technology illustrates, decisions about bioconfinement will take place in a social and ethical context that is not framed solely in terms of quantifiable risks (Thompson, 1997).

Despite its influence on regulatory policy and international discussions, consequentialism does not account for all of the ethical issues raised about genetic engineering (Thompson, 1997) or specifically bioconfinement. Early confinement strategies and the system for their regulation that were developed in the early 1970s focused on a narrower set of concerns that were best addressed by the scientific community. This approach was and remains appealing: the expectation is that clear, rigorous, and concise characterization of existing information about risks, costs, and benefits will lead to informed and acceptable regulatory decisions (NRC, 1996). Nonetheless, the narrow focus ignored potential ethical problems (Wright, 1994) and current approaches continue to ignore moral considerations in favor of science-based issues and utilitarianism (Saner, 2001).

It is useful to recognize that the products of genetic engineering can be organisms that can have social significance or be infused with value for any number of reasons. For example, trees have important symbolic, social, and aesthetic values that present important challenges for biotechnology (McQuillan, 2000; 2001). The desire for "natural" (and even "wild") forests would require stricter confinement methods than simple safety considerations would dictate. To ensure the purity of culturally significant crops, stringent confinement could be necessary.

Beyond risk assessment, there is also the perspective that broader social and ethical values should be considered in determining how much and which methods of confinement would be necessary for various organisms. To address different value systems in bioconfinement decisions, consideration would be given to who is exposed, who benefits, and who decides. RAFI was effective in drawing the public's attention to these questions in the terminator case. The values of specific groups and communities could emerge as important considerations in the choice of biological and physical confinement methods.

A consequentialist framework cannot accommodate the diversity of values in the debate. As a result, the framework can fail to incorporate much of the public's concern about genetic engineering and bioconfinement. Gaskell and colleagues (2002) draw on extensive data collected in the mid- to late-1990s about attitudes toward genetic technologies in the United States and Europe (Durant et al., 1998). There is a striking pattern in the responses that emerged about the risks, benefits, and moral acceptability of genetic engineering that led the authors to conclude "respondents with concerns about gene technology tended to think principally in terms of moral acceptability rather than risk" (Durant et al., 1998).

Public acceptance of bioconfinement methods for GEOs will depend on many of the same factors that influence the public's acceptance of genetic engineering and its products. For bioconfinement to gain public acceptance it will be imperative that lessons are learned from the successes and failures of the past (Eichenwald et al., 2001). In addition to the science-based evaluation used to determine what, if any, confinement strategy should be applied to a GEO, the clear and public articulation of potential ethical concerns is likely to promote public trust and acceptance. Ultimately, the benefits of bioconfinement must be considered. The methods can be powerful tools when combined with other physical, temporal, and biological measures to ensure that GEOs will not harm ecosystems or threaten the food supply.

2

When and Why to Consider Bioconfinement

INTRODUCTION

Safety is an issue for all modern technology—from the design of automobiles to the delivery of information over the Internet—and confinement of technology is frequently an important aspect of ensuring safety. Airbags and radial tires promote drivers' safety; highly developed software promotes computer users' privacy and security. The atomic energy industry works to confine radiation; the chemical industry, to confine toxic chemicals; the food industry, to prevent nonedible industrial rapeseed from contaminating canola oil for human consumption; and the horticultural industry, to confine ornamental plants that might become invasive. The stewardship of those industries bolsters public confidence in technology and its products at the same time as it prevents the damage that can result from the movement of products and byproducts into arenas for which they are not intended. The success of any new technology depends on this attention to safety and confinement and on building public trust by protecting human health, the environment, and the security of intellectual and financial information.

Biotechnology should enjoy the same benefits from attention to safety and confinement. As most traditionally improved organisms pose few safety problems that require confinement, it is likely that most of the vast array of proposed genetically engineered organisms (GEOs) will pose little threat to public health or the environment, and they will require minimal confinement, if any. However, as some traditionally improved organisms require confinement, some GEOs will need some or substantial confinement. If

29

field release of a GEO is proposed, it is important to consider whether confinement is necessary, and, if so, how to attain it.

This chapter reviews the concept of risk and then reviews some of the effects that could be expected from the release of GEOs. Avoiding or minimizing damage generally requires that specific methods of confinement be considered and that the consequences of failure (inadequate confinement performance) are predicted. This chapter also includes a discussion of who bears responsibility for deciding whether and how to confine GEOs and their engineered genes.

WHAT IS RISK?

In many aspects of its deliberations, the Committee on the Biological Confinement of Genetically Engineered Organisms found it necessary to discuss "hazard," "harm," and "risk." The terminology used here is consistent with past reports of the National Research Council (NRC, 1983, 1996, 2002b). To quote a recent report (NRC, 2002b), "as set forth by the NRC (1983, 1996), a hazard is an act or phenomenon that has the potential to produce harm, and risk is the likelihood of harm resulting from exposure to the hazard ... risk is the product of two probabilities: the probability of exposure and the conditional probability of harm, given that exposure has occurred." Risk assessment typically involves asking questions: What can go wrong? How likely is failure? What are the consequences of failure? How likely are those consequences (assuming the triggering event occurs)? How significant are those consequences? How certain are we about this knowledge, whether it is qualitative or quantitative (NRC, 1996)?

Risk assessment involves (1) identifying potential harms, (2) identifying potential hazards that might produce the harms, (3) defining exposure and the likelihood of exposure, (4) quantifying the likelihood of harm given that exposure has occurred, and estimating the severity of the harm (NRC, 2002b). The importance of social values to this determination is discussed below.

Risk is sometimes described as a formula: exposure multiplied by hazard. Although this can be useful shorthand, risk is not easily estimated. Uncertainties can arise from random events (including human error) in the physical world, lack of knowledge about the physical world, or lack of knowledge about the applicability of risk-generating processes (NRC, 1996). Ecological harm (consequences) is difficult to quantify, including damage to an ecosystem or the extinction of a species, for example. Moreover, evaluating harm requires consideration of social values that will define the significance of predicted consequences. Clearly, exact quantification is impossible.

An evaluation of risk must consider cumulative risk—the combined risks to human health or the environment posed by exposure to multiple agents or stressors (US EPA, 2003a). Without cumulative risk analysis, it is impossible to assess the hazard associated with bioconfinement failure. For example, suppose failure would result in the release of a contaminant into the environment. The hazard stemming from that failure could depend on whether the environment is totally free of the contaminant, the contaminant already is present in the environment at a near-hazardous concentration, the environment already is stressed by other factors such that its normal resilience is compromised, or there are some other substances in the environment that would neutralize the contaminant or exacerbate its effects.

The risk of any new technology should be considered in the context of preexisting relevant technologies. Doing so would likely involve an assessment of relative risks and while this comparison is an important one for consideration, it is also a challenging task. In the context of this discussion, relative risk is, in theory, equal to the probability of harm utilizing the new technology divided by the probability of harm utilizing the preexisting technology. One challenge is that, as indicated above, risk assessment is not readily susceptible to exact mathematical calculation. Another challenge is identifying the appropriate preexisting technology to serve as a comparator. Given the diversity of GEOs and the farming systems where they might be used, there is a substantial amount of disagreement as to what preexisting agricultural technologies are appropriate to serve as comparators. Furthermore, given the range and number of potential variables and risks associated with the application of the technologies, identifying the relevant risks for comparison would have to be done on a case-by-case basis. For the bioconfinement technologies described in this report that have yet to be fully developed or applied, discussion of relative risks is simply premature at this time. Nonetheless, as developers of GEOs and the confinement methods progress such comparisons will be attempted but will not be easy to make.

Given the uncertainty involved in risk analysis and the fact that many variables cannot be quantified, the committee determined that an alternate, less formulaic, model for risk assessment would be valuable. Figure 2-1 is a risk assessment matrix that assumes that a hazardous event has occurred.. The hazard of greatest concern is that with a high risk (probability of occurrence) and high significance or severity of harm (black area). Social, ecological, and economic considerations influence the significance of consequences. Depending on the quality of information available, the axes could consist of continuous values or more discrete categories (e.g., a 3×3 matrix of high–medium–low rankings; Miller et al., in press).

Earlier reports of the National Research Council contain lengthy discussions of risk and approaches to its analysis. There is extensive consideration of risk in the report *Understanding Risk* (NRC, 1996), and *Environmental Effects of Transgenic Plants* (NRC, 2002a) contains a detailed

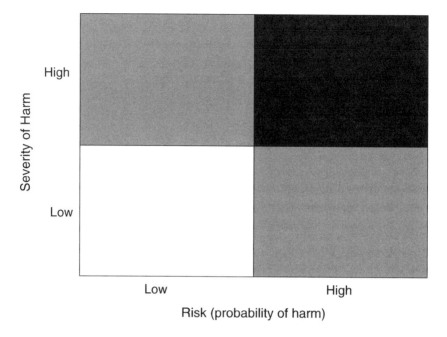

FIGURE 2-1 A risk assessment matrix. The horizontal axis denotes the risk, that is, the probability of occurrence of a harm, and the vertical axis denotes the severity of the harmful consequence.

discussion of risk analysis, including analyses of two-part and whole-organism models, and fault-tree and event-tree approaches. Discussions of risk often distinguish among three related processes: risk assessment, risk management, and risk characterization. The 1996 report sets forth an elaborate description of risk characterization, which it defines as "a synthesis and summary of information about a potentially hazardous situation that addresses the needs and interests of decision makers and of interested and affected parties and which is a prelude to decision making and depends on an iterative, analytic–deliberative process" (NRC, 1996). The committee recognizes those earlier discussions and suggests it would be useful to describe a systematic approach to risk assessment and management, without delving into the issue of risk characterization other than to note its relevance and importance, particularly with respect to transparency and public participation.

Table 2-1 presents a systematic approach to risk assessment and management. It considers exposure, hazard, risk reduction and prevention,

TABLE 2-1 Systematic Risk Assessment and Management

Step	Key questions
Hazard identification	What event posing harmful consequences could occur?
Risk analysis	How likely is the hazard?
	What would be the harms from realization of the hazard, and how severe are they, taking into account social values?
	What is the risk assessment as shown on a matrix of risk (likelihood of harm) plotted against severity of harm; see Figure 2-1, above)? Each cell of the matrix should be accompanied by a qualitative assessment of the response and a quantification of assurance needed to reduce harm if the cell's conditions were to occur.
	How well established is the knowledge used to identify the hazard, estimate its risk, and predict harms?
Risk reduction planning and implementation	What can be done (including bioconfinement and other confinement) to reduce risk, either by reducing the likelihood or mitigating the potential harms? Are there steps that can be taken to prepare for remediation?
Risk tracking (monitoring)	How effective are the implemented measures for risk reduction?
	Are they as good as, better than, or worse than planned?
	What follow-up, corrective action, or intervention will be pursued if findings are unacceptable?
	Did the intervention adequately resolve the concern?
Remedial action	What remedial action should be taken?
Transparency and public participation	How transparent should the entire process be? How much and what type of participation should there be in the steps above (and in risk characterization) by the public at large, by experts, and by interested and affected parties?

SOURCE: Adapted from Kapuscinski, 2001.

monitoring, remedial action, transparency, and public participation (adapted from Kapuscinski, 2001).

Confinement should be undertaken in the context of an integrated confinement system (ICS)—one that considers whether confinement might be necessary from the beginning of GEO development and that has redundancy of confinement as an operating principle (Chapter 6). Elements of this approach include a clear strategy that is adopted in advance of field testing or release of the GEO. Adequate training of those who are responsible for managing the confinement strategy is essential. Permanent employees should maintain accountability and consistency of standard operating procedures. For pharmaceutical-producing GEOs, good management practices also will be necessary. An ICS should be subjected to periodic audit, review, adjustment, and reporting. Each ICS should be supported by a rigorous, comprehensive, and credible regulatory regime that includes mechanisms for inspection and enforcement.

The risk assessment techniques discussed here would have been effective in predicting some notable events that have already occurred. Consider the StarLink incident (Box 2-1). In this case a variety of corn was deregu-

---

BOX 2-1
Confinement Failure: StarLink Corn

The StarLink incident illustrates the failure of a regulatory-agency–industry attempt to prevent genetically engineered material from entering the human food supply. The story offers lessons about the confinement of engineered genes and their products.

Genes from *Bt* that produce various Cry proteins have been individually engineered into plants to confer resistance to insect pests. These are commonly called *Bt* crops. The most widespread use of genes for Cry proteins has been to confer resistance to lepidopteran pests in maize.

The trade name StarLink was applied to maize varieties produced by Aventis CropScience that were engineered to synthesize the Cry9 protein for resistance to the European corn borer. The Cry9 protein was found to be more heat resistant than other Cry proteins and potentially harder to digest based on in vitro studies, suggesting possible allergenicity (US EPA, 1999). Because of the inconclusive tests regarding possible allergic reactions in humans, the US EPA granted StarLink varieties a "split registration" in 1998, designating it for animal feed and industrial use, but not for human consumption (US EPA, 1998).

In September 2000, the *Washington Post* reported that traces from transgenes of StarLink corn had been detected in the human food supply by an independent laboratory for Genetically Engineered Food Alert, a coalition of environmental and food safety organizations (Kaufman, 2000). Soon thereafter StarLink *Bt* gene was found in a great variety of products intended for human consumption, in the United States as well as in Japan and South Korea—major United States maize buyers. The effect on the United States food industry was substantial, ranging from the recall of food products to the temporary shutdown of grain mills (Taylor and Tick, 2003). Aventis voluntarily withdrew its registration, and to prevent further mixing of StarLink material into the human food supply, it agreed to a buyback program with USDA. The US EPA announced that it would no longer grant split registrations for GEOs (Taylor and Tick, 2003).

lated with the proviso that it was not permitted to enter the human food supply, and yet it did so. If the following factors were initially considered—the possibility of pollen dispersal, retention of prior genotypes in the soil seed bank, and the mixing of corn seeds in facilities that harvest, transport, store, and process corn seed—it would have been obvious that without very stringent procedures confinement failure was inevitable.

CONCERNS

The committee adopted the definition of "concern" used in the report *Animal Biotechnology: Science-based Concerns* (NRC, 2002b). "Concern" is used here to mean an "uneasy state of interest, uncertainty, and appre-

The potential sources of the widespread presence of StarLink material in the human food supply are many. Opportunities for mixing genes, seeds, and maize products start in the field and end at the store. Cleaning the facilities that harvest, transport, and store maize seed does not necessarily remove every seed. Processing also presents an opportunity to commingle products. If a farmer does not rotate crops, ungerminated seed from an earlier planting or seed that dropped from another year's crop could germinate and mix with the current year's variety. StarLink genes also were found in varieties into which it had not been engineered, strongly suggesting unintended cross-pollination from StarLink maize. It also is clear that some farmers did not comply with the terms of US EPA registration: Some admitted that they had sold their maize for human food, others stated that they did not know how their maize would be used when they sold it (Pollack, 2001).

Although StarLink corn caused no proven adverse human health effects, other effects have been substantial. Despite the fact that StarLink represented less than 1% of the U.S. maize crop, millions of dollars were spent to buy back grain, consumer confidence in GEOs was shaken, and international trade relationships became strained. Once a gene escapes into populations for which it is unintended, the consequences can be enormous. On December 27, 2002, more than two years after Aventis stopped selling StarLink seed, the *Washington Post* reported that traces of StarLink material were detected in a shipment of U.S. maize on its way to Tokyo for use in food products (Fabi, 2002).

More extensive information on the StarLink incident is presented in *Post-Market Oversight of Biotech Foods* (Taylor and Tick, 2003), which includes a substantial chronology of the incident. Each of the following publications concentrates on a different aspect of the incident: *The StarLink™ Situation* (Harl et al., 2001), *StarLink: Impacts on the United States Corn Market and World Trade* (Lin et al., 2001), and *Channeling, Identity Preservation and the Value Chain: Lessons from the Recent Problems with StarLink Corn* (Ginder, 2001).

hension" (NRC, 2002b). Ethical concerns—a subset—are discussed in Chapter 1.

The committee attempted to base its analyses of risk and concern on current natural and social scientific knowledge. As with its consideration of risk, the criteria include the likelihood that a given result will occur; its severity; and its significance, as would be determined at least in some measure by public opinion or societal values.

Potential Effects

Genetically engineered organisms are controversial. Dozens of scholarly reports have identified, reviewed, and evaluated the realized and potential effects of their use, particularly transgenic crops and fish (e.g., Carpenter et al., 2001; Colwell et al., 1985; Cook et al., 2000; Dale et al., 2002; Hails, 2000; Kapuscinski and Hallerman, 1991, 1994; Keeler and Turner, 1990; Marvier, 2001; McHughen, 2000; NRC, 1989b, 2000, 2002a, 2002b; Pew Initiative on Food and Biotechnology, 2003; Rissler and Mellon, 1996; Scientists' Working Group on Biosafety, 1998; Snow and Moran-Palma, 1997; Tiedje et al., 1989; Traynor and Westwood, 1999; Winrock International, 2000; Wolfenbarger and Phifer, 2000). As GEOs are developed and released, their specific effects can be scrutinized.

Current and Future GEOs: A Brief Summary

The first commercially grown transgenic plant—the Flavr Savr™ tomato (Kramer and Redenbaugh, 1994)—was released to the field a little more than a decade ago. Since that time, millions of acres have been planted in genetically engineered crops—mostly in the United States. Soybean, corn (maize), canola (oilseed rape), and cotton are the major species used worldwide, and most of those plants are restricted to three phenotypic classes: herbicide resistance, insect resistance, and viral resistance (James, 2002). Despite the narrow range of crops and phenotypes and despite the fact that most of the acreage is restricted to a few countries, their rapid acceptance is remarkable in the history of agriculture, outpacing even the acceptance of corn hybrids (James, 2002). Not all field-grown plants that are used to produce commercial products have been deregulated. In the United States, some products have been produced from transgenic crops and viruses in field tests that are regulated under the United States Department of Agriculture (USDA) Animal and Plant Health Inspection Service (APHIS) notification and permit processes (NRC, 2002a).

Transgenic animals are largely under development, and some have been commercialized. In a few cases, transgenic mammals have been created to secrete commercial biochemicals (http://www.nexiabiotech.ca/en/01_tech/

01.php). Those genetically engineered animals that have been commercialized have not been released to the field. Some microorganisms have achieved commercial success with applications in medicine and in food production. As yet, they have been used in industrial situations, rather than in situations calling for field release. For example, chymosin is the active enzyme in rennet, which is used in cheese production. The gene for this enzyme has been engineered into bacteria and yeast, and it now contributes substantially to the commercial production of hard cheese in Europe and North America (Mohanty et al., 1999). Prior to genetic engineering, chymosin was obtained exclusively from rennet extracted from the stomachs of calves.

Benefits beyond the commercial success of transgenic crops also have been noted. In the United States, insect resistance and herbicide tolerance sometimes have permitted reduced pesticide use and have increased yields (Benbrook, 2001; Fernandez-Cornejo and McBride, 2002). The gains in yield and reductions in pesticide use have been modest in developed countries but more substantial in developing nations (e.g., Qaim and Zilberman, 2003). Herbicide resistance probably has facilitated the increasing practice of "no-till" agriculture, which replaces mechanical weed control with chemical weed control, resulting in reduced soil erosion—but the details of how and whether herbicide-resistant plants have improved soil quality require more study (Wolfenbarger and Phifer, 2000).

The list of potential benefits attributable to transgenic plants, animals, and microorganisms is long and diverse, and it is difficult to name all of the beneficial phenotypes that have been proposed for transgenic organisms. However, some expected benefits have been discussed widely. One is the dramatic growth enhancement in several lines of transgenic fish that are approaching commercialization (NRC, 2002b). Also, some traits related to coping with abiotic stresses have been targeted, such as drought and salinity tolerance in crops (e.g., Garg et al., 2002) and cold tolerance in fish (e.g., Wang et al., 1995). Some phenotypes have been proposed to make food healthier by increasing its nutritional value, by eliminating allergens and antinutritional factors, and by extending the shelf life of fruits and vegetables. Other phenotypes pertain to nonfood products processed from plants, such as the alteration of the chemical composition of wood fibers from trees to reduce the cost of paper production, and the production of expensive or hard-to-synthesize specialty chemicals, such as pharmaceuticals in transgenic crop plants (Pew Initiative on Food and Biotechnology, 2002), aquatic plants and moss (Spencer, 2003; Wagner, 2003), algae (Mayfield, 2003), and fish (Aquagene, 2003). Viruses and microorganisms that are natural pathogens of insects and other pests could be engineered into more effective agents of biological control. Transgenic plants, microorganisms, and algae have been proposed for use in environmental remediation to detoxify polluted soils and waters (Cai et al., 1999).

Some of the proposed uses are novel; others are natural extensions of current use. Nontransgenic bacteria already are used to degrade environmental pollutants. Table 2-2 lists some GEOs and phenotypes that have been proposed, are under development, or are in use. Only time will reveal which predicted benefits of those organisms are realized, and a thorough discussion of those benefits is beyond the scope of this report.

Gene Dispersal and Persistence

Before considering specific categories of GEOs that may be candidates for bioconfinement, it is useful to examine how genes and organisms move, and whether transgenes are likely to be favored by natural selection. The ecological spread of transgenes is accomplished by horizontal transfer, dispersal of whole organisms, or dispersal of gametes that may move transgenes past the point of intentional release and into new environments and organisms. The persistence of transgenes depends largely on how they affect the organisms' evolutionary fitness, as discussed below.

How Transgenes Disperse

Horizontal transfer occurs when genetic material from one organism becomes incorporated into an unrelated organism (NRC, 2002a); it is naturally occurring genetic engineering. Gene transfer among unrelated bacteria is relatively common (Syvanen, 2002); transfer among unrelated eukaryotes happens less frequently. The rate of horizontal transfer that is detected as evolutionary change is extremely low—it is far less the mutation rate—although it is surprisingly fast over evolutionary time. And some plants, for example, appear to have acquired genes from other types of organisms, even from other kingdoms (Adams et al., 1998). And some flowering plants apparently have acquired mitochondrial genes from fungi hundreds of times over the past 100 million years (Palmer et al., 2000).

Horizontal transfer is largely viewed as a source of unanticipated consequences. Transgenic plant DNA can transform bacteria in sterile soil microcosms (Nielsen et al., 2000). There is no data yet, however, to suggest that the rate of horizontal transfer would be any faster for transgenic organisms (Syvanen, 2002). But relevant data remain few (Nielsen et al., 1998); further research to identify the mechanisms and ecological implications of horizontal gene transfer (Traavik, 2002) would be helpful.

Dispersal of whole organisms includes the movement of juvenile and adult animals, fertilized eggs and seeds, spores, and vegetative propagules such as offshoots and fragments of plants and algae. Dispersal can occur passively—in the same ways pollen and seeds are transported, for example—or by anthropogenic means: during international trade, in the ballast water

TABLE 2-2 Genetically Engineered Organisms

Species	Engineered Trait	Application	Development
Finfish			
Mud loach	Increased growth rate, improved feed conversion, and likely sterility after insertion of mud loach growth hormone driven by mud loach β-actin regulatory region (Nam et al., 2001a,b)	Aquaculture for human food	Research
Channel catfish	Enhanced bacterial resistance after insertion of moth peptide antibiotic, cecropin B gene (Dunham et al., 2002)	Aquaculture for human food	Research
Medaka	Facilitation of better detection of mutations (presumably caused by environmental pollution) after insertion of a bacteriophage vector (serves as a mutational target). After exposure to mutagenic agent, vector DNA is removed and inserted into indicator bacteria where mutant genes can be easily measured (Winn, 2001a, b; Winn et al., 1995, 2000, 2001)	Industrial uses; environmental uses	Research; method has been patented
Atlantic salmon	Increased growth rate and food conversion efficiency after insertion of Chinook salmon growth hormone gene that operates year-round, thereby fostering steady growth through the year rather than summer growth (Cook et al., 2000; Hew and Fletcher, 1996)	Aquaculture for human food	Method has been patented; seeking Food and Drug Administration approval
Red sea bream	Increased growth rates after insertion of an "all fish" growth hormone consisting of ocean pout antifreeze protein gene promoter and Chinook salmon growth hormone (Zhang et al., 1998)	Aquaculture for human food	Research

continued

TABLE 2-2 Continued

Species	Engineered Trait	Application	Development
Rainbow trout	Improved carbohydrate metabolism after insertion of human glucose transporter type I and rat hexokinase type II, cloned with viral (CMV) and piscine (sockeye salmon metallothionein-B and histone 3) promoters. Potentially allows giving fish feed that contains plant materials (Pitkänen et al., 1999)	Aquaculture for human food; Industrial uses	Research
Steelhead trout	Increased growth rate and food conversion efficiency via insertion of sockeye salmon growth hormone gene (Devlin et al., 2001)	Aquaculture for human food	In use as a research model
Zebrafish	Production of male-only offspring by injecting into fish eggs an altered gene that prevents aromatase enzyme from transforming reproductive hormone androgen into estrogen; lack of estrogen prevents development of female fish (Woody, 2002)	Biological control of aquatic nuisance species, such as carp	Research; In use as a research model
Carp	Improved disease resistance after insertion of a human interferon gene (Zhu, 2001)	Aquaculture for human food	Research
Goldfish	Increased cold tolerance after insertion of ocean pout antifreeze protein gene (Wang et al., 1995)	Aquaculture for human food	Research
Tilapia	Increased growth rate and food conversion efficiency after insertion of tilapia growth hormone gene (Martinez et al., 2000)	Aquaculture for human food	Seeking regulatory approval
Tilapia	Production of clotting factor after insertion of human gene for clotting factor VII, for medicinal applications (Aquagene, 2003)	Pharmaceutical production	Research

TABLE 2-2 Continued

Species	Engineered Trait	Application	Development
Tilapia	Increased growth rate, food conversion efficiency, and utilization of protein after insertion of chinook salmon growth hormone with ocean pout antifreeze promoter (Rahman et al., 2001)	Aquaculture for human food	Research
Mollusks			
Mollusks	Potential improved disease resistance and growth acceleration in mollusks by harnessing altered genetic material from a virus to introduce foreign DNA (Burns and Chen, 1999).	Aquaculture for human food	Research; method has been patented
Oysters	Improved disease resistance by introduction of retroviral vectors. Researchers are determining most effective method of insertion (Burns and Friedman, 2002; Lu et al., 1996)	Aquaculture for human food	Research
Marine Plants			
Seaweed	Enhanced production of carrageenan or agar (valuable to the food, pharmaceutical, cosmetic industries) after introduction of foreign DNA (Cheney and Duke, 1995)	Industrial uses	Research; method has been patented
Micro algae (*Spirulina*)	Potential improved nutritional and medicinal value of commonly consumed *Spirulina*. Method to achieve such trait changes recently confirmed via successful integration and expression of a genetically engineered marker gene (Zhang et al., 2001)	Aquaculture for human food	Research

continued

TABLE 2-2 Continued

Species	Engineered Trait	Application	Development
Algae	Enhanced ability to bind heavy metals after successful expression of a foreign class-II metallothionein (chicken MT-II cDNA) (Cai et al., 1999)	Bioremedial use	Research
Marine Microorganisms			
Diatoms	Reduced dependence on light for growth after insertion of human gene for biochemical involved in metabolism of sugar (Zaslavskaia et al., 2001)	Industrial uses	Research
Crustaceans			
Crayfish	Production of transgenic offspring (in crayfish and live-bearing fish) after injection, in parents' gonads, of replication-defective pantropic retroviral vector. Successful transgenic individuals expressed neomycin phosphotransferase gene (neoR) (Sarmasik et al., 2001)	Aquaculture for human food	Research; in use as a research model
Kuruma prawns	Potential improved growth rate through gene insertion. Researchers are currently inserting marker genes to confirm most appropriate method (Preston et al., 2000)	Aquaculture for human food	Research
Terrestrial Plants			
Corn	Production of a glycoprotein, avidin (Hood et al., 1997; NRC, 2002a)	Industrial uses, including medical diagnostic procedures	Field grown commercially under APHIS notification procedure
Yellow crookneck squash	Transformed with coat protein genes of three viruses to gain resistance against them (Schultheis and Walters, 1998)	Agriculture for human food	Deregulated and commercialized

TABLE 2-2 Continued

Species	Engineered Trait	Application	Development
Flax	Sulfonylurea herbicide resistance based on a gene from *Arabidopsis thaliana* (McHughen et al., 1997)	Agriculture for oil and oilseed production	Deregulated and commercialized
Tomato	Flavr Savr^{TM}; slowed ripening by the introduction of an antisense sequence of polygalacturonase gene from tomato (Kramer and Redenbaugh, 1994)	Agriculture for human food	Deregulated, no longer commercially available
Rice	Enhanced β-carotene production from phytoene synthase and lycopene β-cyclase genes both introduced from *Narcissus pseudonarcissus* (Beyer et al., 2002)	Agriculture for human food with enhanced nutritional value	Research
Poplar	Introduced bacterial gene confers resistance to the herbicide glyphosate (Meilan et al., 2002)	Managed forestry	Field tests under APHIS notification
Loblolly pine	Resistance to pine caterpillars (*Dendrolimus punctatus and Crypyothelea formosicola*) conferred by *Bacillus thuringienesis*. CRY1Ac insecticidal protein gene (Tang and Tian, 2003)	Managed forestry	Research
Walnut	Reduced flavonoid content and enhanced adventitious root formation conferred by a walnut antisense chalcone synthase gene (El Euch et al., 1998)	Nut production	Research
Plum	Resistance to plum pox virus using capsid gene (Ravelonandro et al., 1998)	Fruit production	Release permit issued
Bartlett pear	Resistance to fireblight bacterial disease conferred by a synthetic antimicrobial gene (Puterka et al., 2002)	Fruit production	Field tested under APHIS notification (2000–2001)

continued

TABLE 2-2 Continued

Species	Engineered Trait	Application	Development
Apple	Resistance to apple scab disease conferred by antifungal chitinase genes from the biocontrol fungus *Trichoderma atroviride* (Bolar et al., 2001)	Fruit production	Field tested under APHIS notification (1998–2002)
Papaya	Resistance to papaya ring spot Virus conferred by coat protein gene (Cheng et al., 1996)	Fruit production	Deregulated for commercial use
Banana	Resistance to banana leaf spot disease conferred by the frog *Xenopus laevis* gene for the antimicrobial peptide magainin (Chakrabarti et al., 2003)	Fruit production	Research
Microbes			
Pseudomonas putida	Improved protection against soil-borne pathogens after insertion of genes for production of antifungal compound phenazine-1-carboxylic acid or the antifungal and antibacterial compound 2,4-diacetylpholoroglucinol (Bakker et al., 2002; Glandorf et al., 200a)	Biocontrol of soil-borne pathogens	Small-scale field test
Pseudomonas fluorescens	Chromosomal insertion of two reporter gene cassettes (lacZY and Kan^r-xylE) (De Leij et al., 1995)	To identify effects on indigenous microbial populations in wheat	Small-scale field test
Fluorescent pseudomonads from wheat rhizosphere	Chromosomal insertion of lacZ and Kan^r (De Leij et al., 1995)	To monitor movement of genetically engineered bacteria in soil	Field testing
Agrobacterium tumefaciens K1026	A transfer-deficient mutant of the natural isolate (K84) was constructed that prevents transfer of the plasmid conferring resistance to toxin (Jones et al., 1988)	Biocontrol of crown gall disease	Commercial use

TABLE 2-2 Continued

Species	Engineered Trait	Application	Development
Metarhizobium anisopliae	Addition of *gfp* gene and/or additional protease genes (Hu and St. Leger, 2002)	Biocontrol of insect pathogens of plants	Field testing
Pseudomonas syringae	Ice nucleation-negative (ice minus) mutants constructed (Wilson and Lindow, 1993)	Prevent frost injury to plants	Research
Aspergillus niger	Production of bovine chymosin	Cheesemaking	Commercial
Colletotrichum coccodes	Virulence to weeds	Mycoherbicide	Research and development
Escherichia coli	Production of human insulin by cloned gene	Treatment of diabetes in humans	Commercial production
Insects			
Cochliomyia hominovorax	GFP in PiggyBac (transposon that inserts gene sequence of a desired trait in the TTAA gene sequence of the insect) (Handler and Allen, unpublished)	Improved pest control	Research
Culex pipiens	GFP in Hermes transposon (Allen et al., submitted)	Reduced vector competence	Research
Aedes aegypti	GFP and rescue of eye pigment pathways in Hermes, Mariner, piggyBac (Coates et al., 1998; Jasinskiene et al., 1998)	Reduced vector competence	Research
Tribolium sp.	Fluorescent proteins, "informational molecules" in PiggyBac (Berghammer et al., 1999)	Genetic research	Research
Bombyx mori	GFP, human collagen in PiggyBac (Tamura et al., 2000)	Improved or modified silk, disease resistance	Application in laboratory

continued

TABLE 2-2 Continued

Species	Engineered Trait	Application	Development
Anopheles gambiae	Fluorescent proteins in PiggyBac (Benedict, unpublished)	Reduced vector competence	Research
Anopheles stephensi	Eye color mutant rescue in Minos transposon (Catteruccia et al., 2000; Ito et al., 2002)	Reduced vector competence	Research
Pectinophora gossypiella	Fluorescent proteins in PiggyBac (Peloquin et al., 2000)	Improved pest control, heterologous protein expression in mass-reared insects	Research and field trials
Anastrepha suspensa	Fluorescent proteins in PiggyBac (Handler and Harrell, 2001)	Improved pest control	Research
Musca domestica	Fluorescent proteins in PiggyBac (Hediger et al., 2001)	Genetic research	Research
Ceratitis capitata	White eye mutant rescues, Fluorescent proteins in PiggyBac, Minos (Handler et al., 1998; Loukeris et al., 1995)	Improved pest control	Research

of commercial ships, and in unintentional spilling of seed during transport from harvest to market (NRC, 2002a; Scientists' Working Group on Biosafety, 1998). In the United Kingdom, some roadside feral oilseed rape populations (*Brassica napus*) apparently are replenished by seed that spills from vehicles on their way to an oilseed crushing plant (Crawley and Brown, 1995).

It is difficult to quantify dispersal of whole organisms in detail. Nonetheless, dispersal can cover remarkable distances. For example, bird watchers in North America annually report sightings of dozens of individuals that are otherwise native to Europe. Hundreds of invasive species have successfully colonized new regions after unintentional and deliberate anthropogenic dispersal over hundreds or even thousands of miles (e.g., Mack and Eisenberg, 2002; Rosenfield and Mann, 1992; Williams and Meffe, 2000).

Gamete dispersal provides an opportunity for the sexual transfer of transgenes to wild or domesticated relatives of the transgenic organism. For example, in the case of crop plants, partially or fully sexually compatible relatives would include other varieties of that crop, related crops, and wild relatives (NRC, 2002a). Virtually all farmed fish and shellfish lines can breed readily with other captive lines and with wild relatives. Hybridization of closely related fish and shellfish species is relatively common (Collares-Pereira, 1987; Turner, 1984); it occurs in at least 56 families (Lagler et al., 1977), and it frequently yields fertile hybrids. Motile gametes are typically associated with male function, such as in the sperm cells of pollen produced by seed plants and in animal sperm. The means by which gametes disperse varies by organism. For example, for insects and some vertebrates, sperm are delivered to the female by insemination. But, for seed plants and many aquatic animals (fish and shellfish), gametes are released independently of the paternal parent and either contact an egg under their own power, or require a vector. Wind and insects are agents that most often carry pollen from one seed plant to another; water often serves as the vector for marine and freshwater organisms that release their gametes. Some fraction of gametes released into the environment is expected to disperse, regardless of whether the species is largely outcrossing or mostly self-fertilizing. Bread wheat plants are highly self-fertilizing but can mate with plants some distance away (Hucl and Matus-Cádiz, 2001). Although some crops typically are harvested before they flower, occasionally the plants flower prematurely (Longden, 1993) or are missed by harvesting equipment, and eventually flower. Only a very few domesticated plants—such as some potato varieties and some ornamental plants—are completely male-sterile and produce no pollen.

The dispersal of gametes leading to successful fertilization rarely has been measured directly. The best data, which come from numerous experimental and descriptive studies of plants, show that it is not unusual for 1% or more of the seeds in a population to be sired by plants 100 m or even 1000 m distant (reviews by Ellstrand, 1992, 2003a list several of those studies).

Evolutionary Persistence of Transgenes

When left to their own dispersal devices, organisms, their genes, or both can move into locations for which they were not intended and then multiply. Population genetics theory can be applied to predict the fate of transgenes because the consequences of introducing a new transgene are essentially the same as are those for any new immigrant allele. An immigrant allele's fate—whether its frequency increases, decreases, or stays the same—depends on several factors, the most important of which are the evolutionary fitness effects of that allele in its new population and the rate of recurring immigration (Slatkin, 1987). Other factors that could affect an immigrant allele's expression in a population include counterbalancing gene flow from other sources, changes over time in the environment that alter the allele's influence on fitness, chance effects if the allele is introduced in a very low frequency, and if the population into which it is introduced is small (fewer than 100).

Generally, if an allele confers a fitness advantage when introduced into a population, it is expected to increase in frequency, even if it is introduced only once (Wright, 1969). If an allele has no effect on fitness, and if it is introduced just once, its frequency will be static. That is, if a single immigration event results in a 10% frequency of the new allele, that frequency would be maintained indefinitely. If the neutral allele is introduced repeatedly, however, its frequency eventually should evolve to match that of the source population: If the allele frequency in the source population is 100%, the frequency in the sink population eventually should be the same (Wright, 1969).

Finally, if an allele is detrimental to an individual's fitness, the allele will go extinct in the new population if there is just a single instance of immigration. "Detrimental" is meant to apply to generation-to-generation demographic contributions of a genotype and not to any individual fitness component. If immigration is recurrent, the allele will remain as a polymorphism in the population, maintained by a balance between selection and gene flow (Wright, 1969). The factors that might affect this result are the same as those that determine fitness effects. A basic understanding of gene dispersal and the fitness effects of specific transgenes is essential for evaluating whether bioconfinement is needed and for developing effective bioconfinement methods.

The committee intentionally did not hazard to guess what the fitness effects of classes of transgenes might be in recipient populations because it is already clear that generalizations might be difficult to obtain. For example, experimental field studies have already shown that different pest-resistance transgenes introgressed into wild sunflower have drastically different fitness impacts (Burke and Rieseberg, 2003; Snow et al., 2003). Furthermore, these fitness impacts appear to vary with biotic and abiotic factors.

Some Concerns about Field-Released GEOs

Although most GEOs are expected to carry little or no risks to human health or the environment, several categories of potential risk have been identified by United States regulatory agencies and others. An understanding of the potential problems is relevant to the efficacy of bioconfinement methods that are intended to alleviate them. The discussion of concerns about the release of transgenic organisms has focused primarily on three broad categories of environmental risk: consequences of movement of a transgene or transgenic organism into a population, into a community of wild species, or into a location for which it was not intended; effects of the transgene protein product on other organisms in the ecosystem, such as engineered plant pesticides that could harm nontarget organisms; and evolution of resistance in targeted pests and pathogens. Those categories can apply to conventionally-improved domesticated plants and animals (NRC, 1989b, 2000, 2002a, b). And in fact any new genotype—transgenic or not—can create concerns that are unique to that genotype (NRC, 2002a, b). The first category obviously invites discussion of the need for confinement, including bioconfinement. Although less obvious, there could be circumstances under which bioconfinement would help reduce the chances of occurrences in the other two categories. The movement of transgenes does not, in itself, constitute a risk (NRC, 1987, 2002a, b); it does however constitute the "exposure" component of a risk if a specific hazard is associated with that spread (NRC, 2002a, b).

Three concerns dominate the discussion about the unintended movement of transgenes. The first is whether transgenes will confer a benefit to the transformed organism itself or to weedy or invasive relatives, resulting in the evolution of weeds that are difficult to control or in the evolution of new invasive lineages that overrun and disrupt natural ecosystems. The second issue is the question of whether the wild relatives of transgenic organisms will suffer an increased risk of extinction because of hybridization with or competition from those organisms. A third issue is whether transgenes will spread to other domesticated varieties and whether this could lead to health, environmental, or regulatory concerns. Those concerns are directly relevant to decisions about bioconfinement.

Weediness or Invasiveness

The most publicized concern associated with transgene dispersal in plants is the evolution of weediness or invasiveness, particularly as a result of the sexual transfer of crop alleles to wild relatives (e.g., Ellstrand, 1988; Goodman and Newell, 1985; NRC, 2002a; Snow and Moran-Palma, 1997). It is not unusual for natural hybridization to occur when domesticated and wild relatives live in proximity (Ellstrand et al., 1999; Rhymer and Simberloff, 1996). In the United States, more than half of the top 20 crops are known to naturally hybridize with their wild relatives (Ellstrand, 2003a). One could imagine that genes engineered to confer pest resistance or otherwise increase fitness (such as herbicide resistance or tolerance to abiotic stresses) could contribute to the evolution of increased weediness, especially if the genes were to escape to an organism that already is a weed (the noxious weed johnsongrass is a close relative of the crop plant sorghum).

The problem is not unique to transgenics; hybridization between conventional crops and their wild relatives is known to have led to the evolution of increased weediness and invasiveness in several cases (Ellstrand, 2003a).

The transgenic organism itself could become an environmental problem if the transgenic traits it expresses alter its ecological performance such that it becomes an invasive or nuisance species. Many crop plants—especially those that have had a long history of domestication—pose little hazard because traits that make them useful to humans also often reduce their ability to establish feral populations either in agroecosystems or in nonagricultural habitats (NRC, 1989b). However, feral and naturalized populations are well known for some crops and domesticated animals, and in some cases those populations could become more problematic as a result of their acquiring new transgenic traits. In the United States, forage grasses, turf grasses, alfalfa, and many horticultural species have established free-living weedy populations. Escaped cats, pigs, dogs, and goats have become "feral and resulted in environmental disruptions" in many parts of the world (NRC, 2002b). In the same way, problems have occurred from fish or shellfish species that escape from aquaculture operations (Bartley et al., 1998; Carlton, 1992; Courtenay and Williams, 1992). Introduced tilapia have displaced native fish in African, Asian, and American aquatic ecosystems (Lever, 1996; Lowe-McConnell, 2000), and fish farm escapees are the putative cause of the upstream spread of two Asian species, black carp and silver carp, in the Mississippi River basin (Naylor et al., 2001). If transgenes confer the ability to overcome factors that limit wild populations, the resultant genotype might be significantly more weedy or invasive than is its nontransgenic progenitor.

The factors that limit the invasiveness of populations are not well understood (e.g., Parker et al., 1999). An allele that confers a fitness advantage will spread through a population, but it will not necessarily result in the evolution of invasiveness. Thus, the mere presence of a transgene should not be taken as certainty that the invasiveness of a population has been altered. Many crops are unlikely to become weedier by the addition of a single trait (Keeler, 1989). In a few cases, however, the consequences might be obvious. The evolution of herbicide resistance in a weed population that previously was controlled by that herbicide will force new consideration of options for its control.

Extinction of Wild Taxa

The spread of one taxon sometimes overwhelms related, locally rare taxa, either by competitive displacement or by hybridization, thus increasing the probability of extinction (e.g., Levin, 2003). The fraction of hybrids produced by the rare population can be so high that the population becomes genetically absorbed into the common species (genetic assimilation; Ellstrand and Elam, 1993). Also, hybrids can suffer from reduced fitness (because of outbreeding depression), and the rare species might be unable to maintain itself. For example, spontaneous hybridization between nontransgenic crops and their wild relatives has been implicated in the disappearance of wild coconuts (Harries, 1995) and in the genetic dilution of California's wild walnut populations (Skinner and Pavlik, 1994). Depressed fitness or local extinction of wild fish populations has resulted from introgressive hybridization between an introduced population, often derived from fish farms or hatchery-stocking programs, and a local, genetically distinct wild population (e.g., Hallerman, 2003; Kapuscinski and Brister, 2001; Utter, 2003). If the intended phenotypic effect of genetic engineering permits a GEO to be grown more closely to a wild relative than previously—for example, because it now can better tolerate an environmental stress (saline soil)—the previously isolated species then would be subject to increased interbreeding, which would increase the probability of extinction of the wild population by hybridization. In many other cases, however, GEOs are not expected to exacerbate problems with the conservation of endangered wild relatives.

Gene Flow to Other Domesticated Organisms

The scientific literature has given scant attention to the risk that attends the movement of transgenes from one managed population to another. Hybridization among different transgenic varieties of the same species can lead to the unintended natural "stacking" of transgenes. This already has happened: Hybridization among three canola varieties—each resistant to a different herbicide—has led to the evolution of triple-herbicide-resistant crop volunteers in Canada that are now more difficult to control than were volunteer plants in the past (Hall et al., 2000). Crops that are engineered to produce pharmaceutical or other industrial compounds can cross-pollinate with the same species grown for human consumption, with the unanticipated result of new chemical components in the human food supply (Ellstrand, 2003b). The same issue could apply in the future to gene flow to edible algae from transgenic algae that are created to produce inedible compounds (Minocha, 2003; Zhang et al., 2001). Given current research and development on transgenic fish used to produce pharmaceuticals (Aquagene, 2003), the scenario could extend to fish species also grown for human consumption.

A recent case illustrates the point: soybeans intended for market were contaminated by volunteer maize that had been transformed to create a pharmaceutical compound (USDA, 2002). Cross-pollination was not necessary for the risk to be realized: a persistent seed bank was sufficient. Another consequence of unintentional movement of transgenes among managed populations is the transfer of transgenes into crops or other organisms that are intended to be "transgene-free," such as crops that are to be certified organic or sold to an international market that prohibits the sale of transgenic products. If zero tolerance for nontransgenic ingredients is required by the marketplace, the presence of transgenes in crops or crop products intended to be transgene-free could pose an economic hardship to the grower. Similar concerns apply to organic aquaculture, which is gaining interest and activity (Brister, 2001; Tacon and Brister, 2002), or to the movement of an animal transgene into a crop intended for consumption by vegetarians. Ethical aspects are discussed in Chapter 1.

EFFECTS ON NONTARGET SPECIES

Bioconfinement methods also could be used to prevent unintentional damage to nontarget species, as could occur when a transgene that is designed to interfere with the growth or viability of a pest species could alter other species nearby. *Bacillus thuringiensis* (*Bt*) corn has been developed to control the European corn borer and the southwest corn borer, and corn plants disperse *Bt* pollen. Reports of the potentially toxic effects of *Bt* corn pollen eaten by monarch butterfly larvae (Losey et al., 1999) captured widespread attention, in part because the butterfly is so well known. The effects of *Bt* pollen on monarch mortality appear to be highly variable, depending on factors such as the density of the *Bt* corn pollen, the *Bt* genotype in the crops, and other environmental conditions (Sears et al., 2001; Wraight et al., 2000). It is now clear that current commercial varieties of *Bt* corn are not particularly toxic to monarch butterflies, but it also is clear that some transgene products could harm organisms not intended for control. Nontarget effects of crop resistance alleles that have naturally introgressed into wild populations have not been well researched (Ellstrand, 2003a). Nonetheless, if transgenes or transgenic organisms designed to be toxic to pests or pathogens move into locations or populations for which they were not intended, they could harm organisms other than the intended pest species.

DELAYING THE EVOLUTION OF RESISTANCE

Insects, weeds, and microbial pathogens frequently have evolved resistance to the controls used against them (Barrett, 1983; Georghiou, 1986; Green et al., 1990). As with conventionally bred domesticates, resistance evolution can occur in pests targeted for control by or associated with GEOs. Although the evolution of resistance is a continuous process, the evolution of resistant pests has been considered a potential environmental hazard of GEOs because more environmentally damaging alternative treatments could be needed for continued control. Insect resistance to transgenic

Bt crops is considered inevitable (NRC, 2000), and the United States Environmental Protection Agency (US EPA) has issued guidelines on the cultivation of transgenic crops, mandating that farmers plant refuges of non-*Bt* crops along with *Bt* crops to prevent or decrease the rate of resistant evolution (US EPA, 1999). In some cases, the bioconfinement methods described in this report could help address needs for resistance management.

FOOD SAFETY AND OTHER ISSUES

Beyond the environmental concerns described above, two more topics have been widely discussed. The first is food safety. Although no adverse health effects have been identified after a decade of commercial production of genetically engineered food crops in the United States, initial general concerns about their consequences for human health have been replaced by specific questions: Will some transgenic products prove allergenic? Will transgenic products that are not intended for human consumption end up in foods? Genetically engineered food crops that produce pharmaceutical and industrial compounds pose a special challenge to ensure that those crops do not commingle with crops of the same species intended for food (Pew Initiative on Food and Biotechnology, 2002). There also has been debate about the social and economic consequences of GEOs. The possibility that GEOs could alleviate hunger in less developed nations by increasing productivity is balanced by the concern that genetically engineered crops—like past advances in agricultural technology (Evenson and Gollin, 2003)—will have complex socioeconomic impacts that benefit some farmers and adversely affect others.

WHEN AND WHY TO CONSIDER BIOCONFINEMENT: THE NEED FOR PREVENTIVE ACTION

It is essential to consider preventive action before a failure occurs and even before confinement techniques are chosen. Prevention typically is less expensive and more effective than is remedial action, and some consequences—death of a human, extinction of a species, destruction of a large ecosystem—cannot be reversed. The choice of confinement technique, and even the decision to proceed with a proposed GEO, should be informed by an analysis of possible preventive actions because the use of confinement is itself a precautionary measure.

The committee concludes that it is essential to consider—from the very beginning of the process of developing a GEO and its possible confinement—the risks and consequences of failure, the means of failure prevention (particularly by bioconfinement), and the potential for postfailure remediation, to determine what, if any, bioconfinement measures to take.

Precisely how to identify what combination of confinement measures to undertake—if any—is addressed above.

HOW MUCH CONFINEMENT IS ENOUGH?

If some type of confinement is deemed necessary, careful consideration should be given to how much will be sufficient. In some cases, appropriate confinement might be obtained by conventional, non-biological, methods. For example, sufficient isolation might be achieved by growing a minor crop far from stands of the same variety and away from populations of wild relatives. That might reduce viable pollen flow to acceptable levels. Other steps would ensure that seed and vegetative propagules did not find an environment where they could become established.

The most stringent confinement is necessary when field releases of GEOs or their transgenes have sufficient potential to create substantial problems. If stringent confinement is to be applied to released organisms, the standard methods of spatial or temporal isolation will not suffice. For example, the standard contamination tolerated by breeders of high-quality "foundation" seed generally is 10^{-3} (NRC, 2000). That purity is attained by spatial and temporal isolation, often in conjunction with the use of border or barrier crops that will interfere with pollen flow (Kelly and George, 1998). Generally a 660-ft buffer is considered sufficient to reduce background contamination of maize fields to 0.10%—often considered acceptable. However, almost one-third of some 300 maize fields in the Corn Belt exhibited background contamination that ranged from 1.5% to 15.6% at that distance (Burris, 2001).

Redundancy of methods is usually necessary to achieve stringent confinement levels. The 2002 APHIS requirements for growing transgenic maize for pharmaceutical compounds in the field call for substantial spatial *and* temporal isolation (USDA, 2002). Those requirements, however, are only for preventing dispersal of genes by pollen. APHIS requires additional methods to prevent gene movement by seed or gene persistence in a soil seed bank.

Maximum confinement will not be necessary for most organisms. GEOs that pose no hazard or whose risk level is so low as to be tolerable would not require containment. The need for confinement might vary with the fitness impact of the allele in question as well as with whether immigration is anticipated to be recurrent or a single event. What if the allele will have no impact on fitness? If the allele is expected to have a neutral impact on fitness but immigration is expected to be recurrent, then the allele is expected to increase, generation by generation. If such an allele is predicted to create a significant hazard in locations for which it was not intended and that recurrent immigration is going to occur, then the most stringent confinement conditions are advised. If immigration would occur once or a few times, then less stringent conditions might be permissible, depending on the acceptable frequency for that allele.

What if the allele would prove detrimental in populations or environments for which it was not intended? If only one or a few immigration events are anticipated, the allele will not persist in a population, and if immigration is recurrent, the allele will be maintained at a frequency between zero and one. In this case, the acceptable allele frequency is important. The fact that the allele confers a fitness disadvantage will result in a decline in the average fitness of the recipient population. The fitness change also could be so severe as to drive the affected population to the point of extinction (e.g., Huxel, 1999; Wolf et al., 2001).

Extinction by hybridization can result from more complicated situations. If an immigrant allele reduces viability but dramatically increases mating success, the antagonistic effects on different fitness components can lead to eventual extinction after a single immigration event (Muir and Howard 1999, 2001, 2002). Whether such local extinction would be a problem or a benefit would depend on the population. For example, an increased extinction risk would be a problem for an endangered species, but a benefit for one considered a noxious invasive.

Clearly, to reduce chances of escape, there is a need for as many cost-effective tools as possible. The appropriate amount of confinement might be possible now, but the method could be so expensive as to preclude its use. Bioconfinement methods offer an opportunity to expand the number and diversity of tools available.

Finally, there could be organisms for which the possible hazard is so great that it would be best never to release the product. A recent document (USDA, 2002) states that some species are "inappropriate for the production of pharmaceuticals" when grown under field conditions. The cited example is the oilseed rape species *Brassica rapa*. Its traits include multiple-year seed dormancy, bee pollination, and sexual compatibility with weed species that could be found in adjacent fields.

NEED FOR BIOCONFINEMENT

What sorts of GEOs might require confinement? Some GEOs may raise environmental concerns, as described throughout this report. When great harm to the environment is probable, confinement is warranted. Food safety and food purity issues also could motivate confinement. Other reasons could be social, ethical, political, or economic. For example, the security of intellectual property has long been recognized as a motivation for preventing the unintended escape or theft of living biological material of value (e.g., Ellstrand, 1989). Transgenic crops that are commercialized in the

United States might be prohibited elsewhere, leading to the need to segregate genetically engineered products for export. In addition, APHIS currently requires strict confinement for GEOs that are field-released under "Notification" and "Permit" as part of its performance standards (USDA, 1997). New, more stringent requirements for organisms that produce products intended for use in pharmaceutical and other industrial compounds were released by APHIS in early 2003 (USDA, 2003).

The StarLink incident of 2000 illustrates the need for effective confinement in maize (see Box 2-1). Although StarLink corn was released for animal consumption only, it rapidly entered the general maize supply and within a year its presence was detected "in nearly one-tenth of 110,000 grain tests performed by United States federal inspectors" (Haslberger, 2001). Generally, it is thought that mixing of seed was responsible for most of the unintended movement. However, the Cry9C protein also was detected in other, supposedly genetically pure, non-StarLink varieties, suggesting that cross-pollination was partly responsible for its spread (Taylor and Tick, 2003). Presence may or may not indicate a high level of contamination. Current detection techniques are quite sensitive, capable of detecting a single contaminating grain out of thousands in a bulk sample. It is highly unlikely that the Cry9C protein produced by this variety poses any kind of risk, but the fact that the gene moved so rapidly demonstrates how quickly unintentional movement can occur. In any case, regulatory agencies have seen that granting approval to genetically engineered plants that are intended solely for animal consumption is inadvisable if other varieties are used in human food production.

PREDICTING THE CONSEQUENCES OF FAILURE

Failure of bioconfinement presents several challenges. The most obvious apply to the unintentional escape of transgenes. The specific problem will be the risk that is intended to be averted by the bioconfinement method. Thus, the problem could be environmental; for example, the creation of a new or more difficult-to-control pest that is developed because of the introduction of a gene that makes an organism more invasive. There also could be effects on human or animal health. A gene introduced into a crop for the production of industrial chemicals might inadvertently move into crops intended for human consumption. The problem could be economic. A patented transgene, considered intellectual property, can be stolen. The problem could be the serious decline, displacement, or extinction of a species with social or cultural significance. One could quantify specific risk as the likelihood of failure and the magnitude of escape. Other problems are subtler.

The bioconfinement phenotype itself could cause havoc in organisms for which it is not intended, as in the following example. Because of the automatic and substantial fitness advantage of asexual genotypes relative to sexual genotypes due to the evolutionary "cost of sex" (e.g., Charlesworth, 1989; Williams, 1975), the application of seed apomixis—an often-touted method of bioconfinement—could result in the rapid spread of an asexual genotype through wild, sexually reproducing populations (see Chapter 3 for a detailed discussion). The consequence could be a drastic reduction in genetic diversity and in the potential for evolutionary response (van Dijk and van Damme, 2000).

The failure itself could result in altered public perception of biosafety, decreasing the credibility of the biotechnology industry or government regulators, or both. The StarLink incident led to the realization that current systems do not provide for the segregation of genetic material. A flurry of attention from the popular media about a bioconfinement method that fails could result in similar public mistrust for methods that are designed to keep transgenes in their place, leading to a loss of public confidence in the food supply and damage to the viability of the biotechnology industry as a whole.

Clearly, the failure of GEO bioconfinement can, in some circumstances, result in substantial consequences. What factors must be considered to predict some of the risks that could result from the failure of bioconfinement?

The consequences of bioconfinement failure must be assessed on a case-by-case basis. A science-based risk analysis of the consequences of bioconfinement failure should consider at least the following factors: the organism involved; the trait or traits that have been introduced into the organism; the genomic, physical, and biotic environments in which the failure could occur and that could experience the effects of a failure; the possible effects on human health; the bioconfinement techniques involved; and the social and behavioral factors that could affect consequences.

The most important factors in determining environmental consequences could be the ecological phenotype and ecological novelty of the GEO. For example, mud loach (*Misgurnus mizolepis*) (Chinese weatherfish) is an aquacultural species native to China and Korea. It has been genetically engineered such that no extraneous DNA was used (by inserting a gene and a promoter that originated from the same species—the mud loach itself; Nam et al., 2001a, 2001b). Hatchlings showed dramatically accelerated growth—at a maximum, 35-fold faster than non-genetically-engineered siblings. The largest hatchling weighed 413 g, and, with a length of 41.5 cm, exceeded the size of 12-year-old normal broodstock (89 g, 28 cm). The time required to attain marketable size (10 g) was 30–50 days after fertilization; nonengineered fish require at least 6 months. There also was significantly improved feed-conversion efficiency, up to 1.9-fold. There appeared to be no gross abnormalities other than the size increase, but most individuals died after their body weight exceeded 400 g. Thus, despite the fact that the transgenic organism has no exogenous genes, it has a novel phenotype that is obviously (ecologically) quite different from that of the original (Nam et al., 2001a, 2001b). The degree of transgenic novelty (the number of genetically engineered changes) and the taxonomic or phylogenetic distance between the host organism and the novel genes are not likely to determine consequences (NRC, 2002b). This is also consistent with a previous NRC report where it was noted that both small and large genetic changes can have significant environmental consequences and that the consequences of biotic novelty are strongly influenced by the "genomic environment, physical environment, and biotic environment" (NRC, 2002a).

The significance of undefined consequences depends in part on social values and the context in which the failure occurs. Potential loss of culturally symbolic varieties or species is often a focus for social action because they represent social or spiritual values. For example, a failure leading to the decline of a species will have greater significance if that species has high symbolic importance in that location (sugar maples in New Hampshire, blue crabs in the Chesapeake Bay, corn to the Hopi Tribe, wild rice (*Zizania*) to the Ojibwa and Menomonii, or salmon to sport and commercial fishing and to Native Americans in the Pacific Northwest) than if it has economic or ecological importance alone. In addition, the decline of a symbolic species is a likely indicator of environmental harm.

WHO DECIDES?

Decisions about bioconfinement involve asking whether bioconfinement measures should be applied in a given case and, if so, which measures should be adopted and whether they should be applied alone or in conjunction with other types of confinement. The decision makers come from the genetic engineering industry, including the GEO developers (including academic and government scientists), related industries (such as the wholesale and retail food industries) that could be affected by an escape or by any resulting loss of public confidence; insurance companies; government regulators; and private citizens who might sue to enforce environmental laws. Decisions about private legal action arising from damage caused by GEOs might indirectly affect whether bioconfinement or other confinement measures are undertaken.

Industry

Bioconfinement should be considered early in the development of a GEO. Whether public or private, for-profit or nonprofit, the enterprise that develops a GEO is in the best position to determine the advisability and viability of bioconfinement because that person or group uniquely possesses the information necessary to conduct such analyses. The enterprise has an interest in achieving regulatory approval for field testing and marketing so that it can avoid negative publicity and to protect itself from liability. The industry has an interest in bioconfinement because the escape of one or a few GEOs could jeopardize the viability of the entire industry. The committee is not aware of any industrywide standards (binding or nonbinding) for bioconfinement.

Related industries that use or depend on GEOs, such as the wholesale and retail food industries, also have an interest in maintaining the safety of their products and the public confidence in that safety. Thus, they have an interest in confinement and, by extension, in bioconfinement. The U.S. food industry, for example, supports strong regulations that would ensure the segregation of pharmaceutical crops from those that enter the U.S. food supply. After the Prodigene incident, the Grocery Manufacturers of America met with senior USDA officials and congressional staff members to call for stricter regulation of pharmaceutical crops (Fox, 2003).

Insurance Companies

To the extent that GEO developers can obtain insurance against the possibility of escape or failure of bioconfinement, the risk of loss shifts to the insurer, who then would have an interest in adequate confinement, including bioconfinement. Apparently, insurance is available for genetic engineering under liability insurance policies, and only a few markets contain specific coverage or exclusions of genetic engineering applications (Epprecht, 1998). There is scant experience with losses involving GEOs. At the same time, societal views about the acceptability and value of GEOs— and even of particular bioconfinement methods (such as the use of terminator genes)—are in flux and can influence the insurance risk (e.g., with respect to product liability). Furthermore, the risk associated with some escapes could be extraordinarily high. It thus would seem that the insurance industry operates in a sea of uncertainty. The committee is unaware of any confinement or bioconfinement requirements imposed by insurance companies individually or collectively.

Government

In the United States, GEOs and associated bioconfinement measures are regulated by a mosaic of laws and agencies. The Coordinated Framework for the Regulation of Biotechnology Products was adopted by federal agencies in 1986 (51 Fed. Reg. 23302 [June 26, 1986]) in response to concerns about how best to provide federal oversight for products of biotechnology. Where those laws apply, they take precedence over the private decisions described above, with the exception of decisions concerning private lawsuits for damages.

The Coordinated Framework is based on two premises. The first, which is consistent with the judgment of several reports (NRC, 1989a,b, 2000, 2002a), is that the potential risks associated with GEOs fall into the same general categories as those established for traditionally bred organisms (but see NRC, 2002b, for another judgment). The second is that the statutes written for non-GEOs should provide an adequate basis for regulating GEOs.

The coordinated framework is intended to provide a harmonized regulatory approach and to ensure the safety of biotechnology research and products by using existing statutory authority and building on agency experience with agricultural, pharmaceutical, and other products developed through traditional techniques of genetic modification. The development of the framework anticipated that agencies might need to develop specific regulations or guidelines under existing statutory authority. It also anticipated institutional evolution in accord with experience, including modifications made through administrative or legislative action. Finally, the coordinated framework specifies that interagency coordination mechanisms are necessary to address all manner of policy and scientific questions (Council on Environmental Quality and the Office of Science and Technology Policy, 2001).

The regulatory approach articulated by the coordinated framework invokes many statutes, as well as their implementing regulations and guidelines that could apply to GEOs. Some apply to specific products or activities and are administered by a single agency; others apply generally and are thus of interest to virtually all agencies. The committee finds that the complexity of federal oversight could hamper the effectiveness of bioconfinement implementation, as discussed below.

Determining which law applies to a GEO under the coordinated framework can entail a complicated analysis, involving such factors as the stage of development (Is the GEO contained in the laboratory? Is it being field tested? Is it ready for commercial use?), its uses (Is it intended for bioremediation of pollution or for biocontrol of another organism? Is it intended to be a human food or drug or an animal biologic? Might it eventually be used as food even though that is not its current use?), the type of possible hazards (Could it harm plants? Could its genetic material cause a plant to become a noxious weed? Could it release pollutants into the atmosphere or water?), the type of organism (Is it an animal, plant, or microorganism?), and whether regulatory agencies have reached consensus on how the GEO should be regulated.

Depending on that analysis, many laws and several agencies could be involved in regulating a particular GEO or bioconfinement method, and several other statutes that are not currently used to regulate GEOs could be brought to bear. Some of them are now applied to invasive species, and the experience with and laws that regulate nonindigenous species might be helpful in regulating GEOs (Council on Environmental Quality and the Office of Science and Technology Policy, 2001).

Federal statutes and guidelines embody different approaches and contain different authorities, standards, and enforcement provisions. Other federal agencies, including the Office of Management and Budget, for example, affect the way regulatory agencies interpret and apply the statutes for which they are responsible. Congress also weighs in by controlling funds. Congress prohibited US EPA from expending any money to enforce a regulation issued under the Clean Air Act. Thus, each agency exercises regulatory authority differently. To the extent that laws regulate confinement and bioconfinement, they trump the decisions of the private parties discussed above.

Several agencies impose specific confinement requirements. APHIS, for example, requires physical confinement where crops genetically engineered to produce industrial chemicals or pharmaceuticals are field tested. The confinement requirements include a 50-ft perimeter fallow zone around the field test site; restriction on the production of food and feed crops in the field test site and in the fallow zone the next season if volunteer plants could be inadvertently harvested with that season's crop; the use of dedication of mechanical planters and harvesters for the duration of the test, and cleaning of that equipment in accordance with protocols for tractors and tillage attachments; the use of dedicated facilities for storing equipment and the regulated GEO; and, for field tests of open-pollinated pharmaceutical corn, a prohibition against growing any other corn one mile of the field test site for the duration of the field test (Federal Register, 2003).

Another example is that the US EPA sometimes requires refuges or buffer zones for certain genetically engineered plants (*Bt* corn and cotton) (US EPA, 2002; Elias, 2002). In 2002, the US EPA filed complaints against two companies for noncompliance with agency requirements not to grow experimental GE corn too close to other crops and to use trees and other corn to form a windblock next to the GE crops. The complaints were settled through legal agreements between EPA and both companies, with each company agreeing to pay a fine (Elias, 2002). Additionally, EPA fined one of those companies a second time, for failing to notify the agency of test results indicating the presence of the experimental gene in seeds grown near the experimental plants, and for failing to submit maps identifying the location of such seeds (EPA, 2003b).

The Food and Drug Administration (FDA) may require confinement (including bioconfinement) of transgenic animals under the terms of the Federal Food, Drug, and Cosmetics Act (FFDCA), 21 U.S.C. 321-397. The

U.S. Fish and Wildlife Service and the National Marine Fisheries Service are permitted by the Endangered Species Act (16 U.S.C. 1531-1544) to require bioconfinement for protecting a listed species.

Laws concerning confidentiality of information also affect how GEOs and their confinement are regulated by the federal government. Trade secrets and confidential commercial information—often called confidential business information (CBI)—are protected under the statutes that are used to regulate GEOs. The CBI provisions in those statutes differ considerably, as do the regulations issued thereunder and the procedures used by the agencies involved. For example, FFDCA prohibits FDA from sharing CBI with any other federal agency, even if that agency is engaged in evaluating or regulating the same GEO. In contrast, other statutes used to regulate GEOs do not contain any such prohibition. Each agency administers its own program for protecting CBI on a statute-by-statute basis. Thus, US EPA has two separate CBI programs, and authorization for access to CBI under one does not allow access under the other. The result is that a scientist involved in regulating a GEO under one statute cannot share CBI, including confinement-related CBI, with a scientist involved in regulating the same GEO under another statute unless each is qualified to know CBI under both statutes. Similarly, each agency applies different amounts of scrutiny to assertions by business entities about what constitutes CBI. CBI is not available to the public (for examples of the use of CBI, see Chapter 3, Tables 3-2 and 3-3).

The regulatory system described above is obviously complex, leading to differing legal authorities and responsibilities, potentially overlapping jurisdictions, and differing statutory standards for regulating GEOs and determining confinement (including bioconfinement). This could lead to a host of problems in regulating confinement. For example, US EPA has no independent authority to require information about where a GEO granted nonregulated status by USDA is grown, which impedes US EPA's ability to monitor confinement requirements for that GEO. US EPA does not have the regulatory authority to enforce the confinement requirements built into the Federal Insecticide, Fungicide, and Rodenticide Act (FIFRA) in any event (Bratspies, 2002). US EPA may set regulatory thresholds for GEOs, but FDA has no such authority (Taylor and Tick, 2003).

Coordination is needed, but there are no central mechanisms to ensure consistency in confinement, including bioconfinement, or to ensure coherence in setting priorities (Bratspies, 2002). The splintered approach to protecting CBI further interferes with agencies' ability to take a coordinated approach to GEO confinement.

Indeed, a mismatch exists between the application of existing laws and the actual practice of regulating GEOs. Because the statutes initially were enacted for radically different purposes, there is no unifying vision of or approach to the questions and challenges posed by GEOs and their confinement and there is no concordance with the coordinated framework's evident goal of smoothing the path for genetic engineering technology (Bratspies, 2002). The approaches normally used in carrying out those statutes do not always fit GEO confinement. For example, US EPA typically applies split registration for traditional chemical pesticides by requiring labeled containers with a clear indication of registered use. This practice cannot be extended to *Bt* crops, for example, because they do not stay in neatly labeled bottles, and the postharvest distribution, storage, and processing of the corn do not parallel the use of a typical pesticide.

Citizen Suits to Enforce Environmental Laws

Private citizens have a role in enforcing relevant legal authority: Some statutes allow them to bring action to enforce environmental laws, either to direct the agency involved to enforce a law or to seek a civil penalty (which, if recovered, would go to the government). Citizens can sue to enforce the National Environmental Policy Act and the Endangered Species Act as they apply to the bioconfinement of GEOs. Citizen suits seeking to compel government agencies to enforce the law are permitted by other environmental statutes including the Clean Water Act (sections 304 & 305, 42 U.S.C. sec. 7604, 33 U.S.C. 1365), and cases frequently are brought under such provisions. Of the statutes used to regulate GEOs, only the Toxic Substances Control Act (TSCA) contains such a provision (U.S.C. sec. 2619). In a related type of action, however, it is possible to petition USDA, FDA, and US EPA, respectively, under the terms of the Plant Protection Act, FFDCA, and the FIFRA regarding some GEO-related activities and to challenge the agencies' decisions regarding those petitions in court under the judicial review provisions of those statutes.

Private Action for Damage

An indirect decision maker about bioconfinement is found in private causes of action for compensation for damage that results from escape, which typically would be subject to state law. One example with respect to nonbiological confinement is a class action lawsuit brought by farmers who claimed they were harmed when StarLink corn was discovered in the human food supply in 2000. The biotechnology companies that created and distributed StarLink agreed to pay $110 million to settle the case (*Pesticide and Toxic Chemical News*, 2003). Private legal action also can be used to enforce or invalidate patent rights, including in the context of a failure of confinement (Monsanto Canada Inc. & Monsanto Co. v. Percy Schmeiser & Schmeiser Enterprises Ltd., 2001 FCT 256, confirmed in Percy Schmeiser & Schmeiser Enterprises Ltd. v. Monsanto Canada Inc. & Monsanto Co., 2002 FCA 309, on appeal to the Canadian Supreme Court as of May 2003). Such law is federal, thus implicating lawmakers at the federal level.

3

Bioconfinement of Plants

METHODS OF BIOCONFINEMENT

Many approaches have been proposed for the biological confinement of plant transgenes (Table 3-1; Daniell, 2002). Some are based on pre-existing agronomic or horticultural methods, others are newly developed, and some are hypothetical. In a few cases, there are data that illustrate the efficacy of those approaches; in other cases, the approaches are untested. This chapter reviews and analyzes as many bioconfinement methods for genetically engineered plants as the committee could identify, although the survey is incomplete because new methods are proposed constantly. The discussion begins with strategies for blocking sexual and vegetative reproduction. Other techniques that reduce the spread and persistence of transgenes in wild and cultivated populations of plants are reviewed. The chapter also considers—as best as possible, given the limited data available—the efficacy of those methods at various spatial scales. There is a discussion of whether the methods could affect the populations and ecosystems in which they are deployed. Given that bioconfinement methods are expected to be less than 100% effective, the chapter also asks how to monitor for escape of plant transgenes and whether detection and subsequent culling would be an effective backup to a primary bioconfinement method. Case studies are provided to highlight the bioconfinement issues specific to transgenic trees, turfgrasses and algae. The chapter concludes by asking what consequences might accrue and what mitigation might be necessary if bioconfinement and monitoring of genetically engineered organisms (GEOs) fail.

TABLE 3-1 Bioconfinement Methods in Plants

Purpose	Method	Major Limitations	Other Considerations
Confine all gene flow *via* pollen and seeds	Sterile triploids or interspecific hybrids	Few triploid or sterile hybrid cases apply or are effective	Not useful if seed production is desired
	Use only male or only female plants that can be propagated vegetatively	Not feasible if same species or compatible relatives could cross-pollinate with unisexual plants; sex expression can be leaky	Not useful if seed production is desired
	V-GURTs, such as original terminator	V-GURTs under development (early); other sterility methods require vegetative propagation	V-GURTs should not be used in food crops if growers need to save seeds
Reduce spread and persistence of vegetative propagules	V-GURTs with inducible promoters that kill vegetative tissues	Under development (early)	
Confine pollen only	Male sterility	Available for some species, could be lost in later generations; transgenic methods could be more durable	Crop requires other plants as source of pollen if seed production is desired
	Transgene in chloroplast; maternal inheritance	Under development; not feasible for plants with paternal inheritance of chloroplast DNA (most gymnosperms)	Possible to obtain high concentrations of desired genetically engineered proteins, but many traits cannot be conferred by chloroplast genes
	Cleistogamy (closed flowers)	Under development (early)	Results in self-pollination
	Apomixis (asexually produced seeds)	Under development (early)	Hybrid varieties would have high yield and breed true; could become invasive

TABLE 3-1 Continued

Purpose	Method	Major Limitations	Other Considerations
Transgenes absent in seeds and pollen	Transgenes only in rootstocks	Under development (early); cannot use transgenic traits in flowers, fruits, seeds	Applicable to grafted scions of certain woody species such as grapes, fruit trees
	Transgenes excised before reproduction	Under development (early); very speculative; cannot use transgenic traits in flowers, fruits, seeds	Allows seed production without spread of transgenes
Confine transgenic traits only (transgenes can spread)	T-GURTs involving inducible traits	Under development (early); external cues for transgene expression might not be reliable enough for high efficacy	Potentially useful; avoids concerns about sterile plants, but inactive transgenes can still spread
Reduce gene flow to and from crop relatives	Repressible seed lethality (see Fig. 3-2)	Under development (early)	Allows viable seeds to be produced on same cultivar. Seeds sired on other cultivars or wild relatives would not be viable
	Cross-incompatibility	Under development (early); speculative	
	Chromosome location in allopolyploids	Under development; possible if relative has nonhomologous chromosomes; can be leaky	Applies only to crops that are allopolyploids (wheat, cotton, canola)
	Tandem constructs to reduce fitness in crop-wild hybrids and their progeny	Under development (early); requires fitness-reducing trait detrimental to wild plants but not crop	

continued

TABLE 3-1 Continued

Purpose	Method	Major Limitations	Other Considerations
Phenotypic and fitness handicaps to reduce need for confinement	Domestication phenotypes	Under development; does not prevent gene flow	
	Auxotrophy (dependence on specific nutrients or growing conditions)	Under development; does not prevent gene flow	
Reduce exposure to transgenic products in plants	Tissue- and organ-specific promoters that limit expression of transgene	Promoters available, but greater efficacy needed in many cases; confines transgenic traits but not the transgenes; transgenes can spread	Could alleviate the need for bioconfinement in some cases
Minimize or eliminate need for bioconfinement	Choice of alternative organisms; choice not to release in field; choice not to proceed with GEO	Economic costs can be high, especially if decision to change course is made after economic investment	Often feasible and highly recommended when appropriate; alternative choices should be examined before GEO is developed

For thorough confinement, pollen dispersal, seed dispersal, and vegetative persistence must be considered. V-GURT, variety genetic use restriction technology; T-GURT, trait genetic use restriction technology.

Sterility

Because transgene escape by pollen or seeds is not possible for plants that do not produce fertile pollen or seeds, the task of bioconfinement is simplified because it is necessary only to keep track of vegetative dispersal units, such as tillers, rhizomes, and stolons. Bananas and seedless grapes are among the sterile food crops that are propagated vegetatively. Many *non-sterile* cultivated plants are sold as cloned vegetative material, including some varieties of potato, turfgrass, and ornamental plants and poplar trees. Several mechanical, chemical, and genetic methods can be used to block the production of fertile pollen or seeds in those plants. This section reviews genetic approaches that achieve sterility. They include nontransgenic methods (triploids); transgenic sterility that is nonreversible; and transgenic approaches that allow for reversible sexual sterility that permits further breeding. The sections that follow discuss options for blocking vegetative spread and for obtaining male sterility.

Interspecific Hybrids

Interspecific hybrids often exhibit partial or full sterility (e.g., Grant, 1981; Stace, 1975). The sterility of the mule, a horse and donkey hybrid, is well known. In some cases, interspecific hybrids have almost complete male and female sterility. However, most interspecific plant hybrids are *not* fully sterile (e.g., Stace, 1975). In a surprising number of cases, hybrid fitness has been shown to be as high as or higher than that of the parental genotypes (Arnold, 1997; Arnold and Hodges, 1995). For example, Arriola and Ellstrand (1997) compared the fitness of hybrids of *Sorghum bicolor* (the crop, grain sorghum) and *S. halepense* (the weed, johnsongrass) and genetically pure *S. halepense* siblings under field conditions. They report that the hybrids did not significantly differ from the weeds in terms of biomass, tiller number, seed set, or pollen viability. Furthermore, in many species, relatively or fully sterile hybrids reproduce and spread by vegetative reproduction, sometimes even more vigorously than do their sexually fertile relatives (e.g., Ellstrand et al., 1996). It is well known that the fitness of hybrids varies tremendously in different environments (Anderson, 1949; Arnold, 1997). Thus, housing transgenes in interspecific hybrids might afford some moderate bioconfinement relative to nonhybrids, but for any given hybrid genotype, male fertility, female fertility, and vegetative reproduction (if appropriate) must be measured in a range of potential field environments to allow an estimate of what amount of bioconfinement might be expected.

Strengths

In cases where there is complete or near-complete sterility, interspecific hybridity could yield a reasonably easy way to obtain bioconfinement in plants, as in the case of triploid hybrids. As long as sterility is maintained in a variety of environments, the genes of those plants are unlikely to spread through pollen or seed.

Weaknesses

Sterile interspecific plant hybrids will not be a general solution for plant bioconfinement. Specific hybrids might prove to be very sterile, but it is more likely that interspecific plant hybrids would offer moderate bioconfinement at best and no bioconfinement at all in some cases.

Sterile Triploids

Breeding methods that disrupt chromosomal pairing during sexual reproduction have been used to create sterile plants. Most plants are chromosomally diploid (characterized as 2n). That is, they have two sets of matching homologous chromosomes in their somatic cells. The two sets pair up and separate during the process of gamete formation, and the number of chromosomes is halved for each pollen grain or ovule (those gametes have n chromosomes). The diploid number is restored when the gametes fuse to create a zygote.

Organisms with three sets of chromosomes are called triploids (3n). In humans, triploidy is lethal, and it is a rare condition in wild organisms (Chapter 4). It is not uncommon in cultivated plants (Grant, 1981), however, many commercial banana cultivars are triploid and thus seedless (Simmonds, 1995). Spontaneous triploids primarily appear to result from the fusion of a normal gamete (n) with an aberrant unreduced (diploid, 2n) gamete. Spontaneous triploids also can occur from the fusion of a gamete from a diploid species with one from a related tetraploid (4n) species (which produces gametes that bear 2n chromosomes). For example, if a 2n plant is crossed with a 4n plant, all of their progeny would be 3n and would be expected to be sterile. Triploid plants found in the wild typically are partially or fully sterile with respect to pollen and seed production. Those that are fully sterile persist only if they are capable of asexual seed production (apomixis) or vegetative reproduction. Triploidy in cultivated plants is maintained mostly through vegetative propagation. Thus, induction of triploidy (and other odd-numbered chromosome counts) represents a possible option for bioconfinement.

Chromosomal situations other than odd ploidy—extra or missing individual chromosomes (aneuploidy) and translocation heterozygosity—also disrupt gamete formation during meiosis. Although they can cause reduced fertility, they apparently have not been examined for use in bioconfinement. More information on chromosomal variation in plants and its consequences for plant fertility is found in Burnham (1962) and Levin (2002).

Strengths

If triploidy results in pollen and seed sterility, and if the degree of sterility does not vary from one environment to another, induction of triploidy could be an effective method of bioconfinement. Triploidy induction will be most effective for organisms that do not reproduce asexually, although that complicates options for further breeding and multiplication. Triploidy also can be induced in other transgenic organisms such as fish (Chapter 4).

Weaknesses

Much like interspecific hybridity, the efficacy of triploidy induction varies by genotype and environment.

Unisexual Plants Lacking Mates

Many dioecious (unisexual) plants can be propagated vegetatively, among them holly, kiwi, gingko, avocado and asparagus, such that only one sex is used for genetic engineering. Sex-specific molecular markers can be used to identify male or female plants before massive propagation (e.g., Khadka et al., 2002; Reamon-Buttner, 1998). In fields, bioconfinement could be achieved if such plants are grown in unisexual stands far from conspecifics or wild relatives with which there could be cross-pollination. For example, all-female cultivars of ornamental nonnative plants could be used in this context. However, this method of bioconfinement is unlikely to be practical in most cases. First, the number of species for which the conditions would be met (along with sufficient economic advantages) is small. Second, dioecy is known to be quite leaky (Krohne et al., 1980; Poppendieck and Petersen, 1999); seeds could be produced in low frequency by "male" plants, especially in large-scale plantings. Finally, human error could result in mix-ups that allow both sexes to occur in the same population, resulting in a breakdown of bioconfinement.

Strengths

This method might be desirable if it is used in combination with other confinement approaches in small-scale plantings.

Weaknesses

This method is unlikely to be reliable, and it applies only to a narrow range of species.

Transgenic Sterility

Transgenic methods are available for developing plants that abort young flower buds and thus become sterile through ablation. The resulting plants cannot be used for breeding or for multiplication by seed, but this method has been considered for some clonally propagated plants, such as poplar trees. Strauss and colleagues (1995) reviewed the rationale for attempting to engineer nonreversible sterility in forest trees. One strategy for creating sterility-causing transgenes that is particularly attractive for perennial plants is to ablate floral tissues by the expression of cytotoxin genes that are fused to developmentally induced promoters expressed in flowers. Promoters from floral-specific genes tend to work well across species. Thus, ablation methods based on these genes probably will not require cloning of new gene homologues from each new transgenic species and genotype. Practical constraints include the requirement for vegetative propagation if complete sterility is engineered and the need for sterility to be highly stable in long-lived species such as trees and perennial grasses. Strauss and colleagues (1995) suggest that long-term stability could require suppression of more than one floral gene or use of more than one genetic mechanism for sterility.

A shortcoming of nonreversible sterility is that it precludes options for further breeding and seed production within the genetically engineered line that could be needed in the future. For trees or other perennials that do not flower in the first 5–10 years—the breeding period is longer than the generation of new transformants—that limitation might not be a major concern because new transformants could be made within the same period. The engineering of sterility by ablation can be conducted as the last step in the improvement process after breeding or genetic engineering for other traits has been accomplished. The preablation, fertile versions of the lines would still be available for use in breeding or seed production.

Reversible Transgenic Sterility

Plants that are permanently sterile, such as those described above, constitute an evolutionary dead-end. Researchers have proposed various transgenic methods by which sterility can be gained or lost by design (Figure 3-1; Daniell, 2002). One type of reversible sterility blocks gene flow through pollen and seeds, thereby, for example, preserving a seed company's ownership of transgenic germplasm. With this method, transgenes that confer desirable traits are linked to transgenes that cause sterility, and the two are inherited together. Because this strategy restricts access to fertile plants, it is known as variety genetic use restriction technology (V-GURT). Trait genetic use restriction technologies (T-GURTs) induce transgenic traits in fertile plants by means of a specific stimulus, such as a chemical spray. The term GURT has gained wide use in scientific and policy discussions (e.g., FAO, 2002), but this report focuses on bioconfinement uses of GURTs and related techniques, keeping in mind that incentives for developing those methods are often based on proprietary commercial goals.

One of the first V-GURTs was the so-called terminator technology protection system in which transgenic plants produced dead seeds. V-GURTs have not yet been used in any deregulated or commercialized crops, but, the terminator technology patent application was extremely

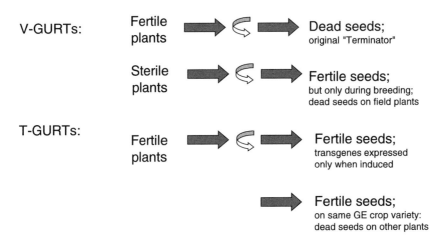

FIGURE 3-1 Proposed transgenic bioconfinement methods in plants.
V-GURT, variety genetic use restriction technology; T-GURT, trait genetic use
restriction technology.

controversial, especially in developing countries. The V-GURT approach
induces seeds that grow into plants that produce nonviable offspring when
they are cultivated in farmers' fields. Induction can occur by soaking the
seed source in a solution that induces a promoter, setting the stage for late-
acting lethality in ripened seeds (Figure 3-1, V-GURT example 1). In field-
grown plants, a promoter that is expressed late in seed development activates a lethal gene that renders the seeds unviable but still fully formed,
which is important if the seed is to be sold for food, feed, or other uses.
However, seeds in the original seed lot that are not induced properly can
develop into fertile plants rather than sterile ones. Such incomplete sterility
seems quite likely, based on the status of the technology (Daniell, 2002),
and other V-GURTs are likely to be more effective. To avoid the problem of
incomplete induction of sterility, plants could be engineered with sterility as
the default condition, and breeders could use a stimulus to induce a promoter to render them fertile (Figure 3-1; adapted from FAO, 2002).

Several related transgenic sterility methods are in development commercially and by independent researchers, but little has been published
about them beyond general descriptions in patent applications (FAO, 2002).
One exception is the research published by a group that developed a method
called "recoverable block of function" (Kuvshinov et al., 2001), which
consists of a DNA sequence element (a "blocker") that interrupts a specific
molecular or physiological function in the host plant, leading to death of

the host plant or its seeds. A second DNA sequence element (for "recovery") restores the blocked function in the host plant. The blocker and the recovery sequences are physically linked to the transgene of interest in one construct so that they integrate into the genome together and remain united during sexual reproduction. The recovery function is designed to be activated by exogenous chemical or physical treatment. Thus, the dispersal of pollen or seeds with the recoverable block of function construct would result in progeny that would die or be unable to reproduce because the recovery function would be inactive. The work is still in the early stages, and it might or might not reach commercial development.

Sterility systems for genetically engineered plants have been criticized because they would prevent growers from saving seed and having the option of using transgenes to improve local varieties. If implemented widely, V-GURTs such as the terminator technology would force growers to buy new seed each year to benefit from modern varieties. Many growers do buy new, certified seed each year, to save time and obtain a high-quality product that is free of contaminating pathogens and weed seeds. Many food crops and annual ornamental plants are sold as F_1 hybrids, among them corn, sunflower, and petunias. Seeds from those plants can be saved but they do not "breed true," so new seeds must be purchased each year. The socio-economic issues surrounding V-GURTs and other sterility methods are discussed in Chapter 1. Environmental effects of the methods are discussed later in this chapter. V-GURT methods could be useful for bioconfinement of grasses, trees, and other horticultural species in which it is desirable to strongly limit gene flow. The social, political, and ethical issues attending the use of V-GURTs in food crops will need to be addressed.

Strengths

Reversible sterility methods could become very useful for bioconfinement because they could be used to block the dispersal of pollen and seeds that bear unwanted transgenes.

Weaknesses

The effectiveness of those novel methods has not been determined nor has their acceptability to consumers. The efficacy of reversible sterility could be diminished by gene silencing or recombination events that cause the sterility construct to become dissociated from the transgenes that require confinement. Research is needed to develop appropriate inducible promoters. Public access to data on the efficacy of transgenic reversible sterility, including long-term studies of transgene stability, will be essential. The technology should not be used in food crops for which growers need to save seeds for future planting or breeding. Possible environmental concerns should be evaluated on a case-by-case basis and are discussed later in this chapter. V-GURTs will not prevent clonal propagation of many plants, such as some species of grasses, shrubs, and trees.

Mortality of Vegetative Propagules

Vegetative spread, both natural and human-mediated, is common in perennial species. Vegetative clones of semidomesticated and nondomesticated grasses, trees, and shrubs can spread over large areas and survive for decades as new ramets are produced and old ones die off. Some plants—especially species that occur along river margins and shorelines—also have vegetative parts that break off and disperse. Many perennial crops, horticultural plants, and woody species can be multiplied and distributed by rooting clonal segments of the plant and meristematic tissue. Depending on the plant's growth habit and ability to be cloned, strategies for minimizing vegetative propagation could be an essential component of bioconfinement. The ability to propagate plants vegetatively is often desirable for commercial production, but in wild species, this trait often is associated with enhanced competitive ability.

Transgenic methods can be used to restrict the spread of vegetative propagules, such as tillers, rhizomes, and root suckers. Given that it will rarely be practical to breed plants that have lost this ability, one of the few options for bioconfinement of vegetative parts is to use a GURT that is induced to kill the plant at some point in its development before it is cloned or propagated (FAO, 2002). Many inducible promoters could be used, including those triggered by chemical applications or winter conditions.

Programmed cell death (PCD) is a normal part of development, and, when it is better understood, that response to stress in plants as well as animals (Zhivotovsky, 2002) could be developed into a transgene bioconfinement method for vegetative propagules. Pontier and colleagues (1999) observed that a senescence-like process is triggered during the formation of necrotic lesions in disease-resistant plants. They suggested that cells committed to die in resistant plants during this hypersensitive response (HR) to pathogens might release a signal that induces senescence in neighboring cells. The signaling pathway responsible for PCD and HR involves changes in the antioxidant systems that are activated by nitric oxide and reactive oxygen species (De Pinto et al., 2002). AtMYB30, transcriptional regulation gene, has been identified as a positive regulator of the hypersensitive cell death program in plants in response to pathogen attack (Vailleau et al., 2002). Several lesion mimic mutants have been isolated in *Arabidopsis* and in other plants that display accelerated HR (Jambunathan et al., 2001). Lesion mimics also can be generated in plants by various transgenes (Mittler and Rizhsky, 2000), such as the lethal leaf spot 1 (Lls1) gene, which is conserved between plant species and acts to suppress cell death (Spassieva and Hille, 2002). A virus-induced silencing of the Lls1 gene in tomato produces a phenotype that resembles the Lls1 mutant in maize, in which large necrotic lesions form in response to aging and environmental stresses, such as light, wounding, and pathogen attack. Those results suggest that, with additional research, it will be possible to environmentally trigger widespread cell death in escaped vegetative propagules of genetically engineered plants either by overexpressing the antioxidant-signaling-pathway genes or by regulating the upstream genes, such as AtMYB30 and LLt1, which control PCD.

Strengths

It is theoretically possible to develop transgenic suicide systems that can be induced to block vegetative reproduction.

Weaknesses

Designing plants that self-destruct reliably at a given time or stage is a formidable technical challenge that could require extensive research and development. Any bioconfinement method that relies on transgenic approaches and inducible promoters could fail because of gene silencing, recombination, or incomplete induction of specific promoters in the transgene construct.

Confining Pollen-Mediated Spread of Transgenes

The discussion thus far has addressed biological methods for confining general reproductive capability in transgenic plants. Confinement of pollen-mediated gene flow can be used to reduce the need for physical isolation of transgenic plants, especially if seed dispersal and vegetative spread are of no concern. This section begins with a discussion of male sterility, which can be achieved through conventional and transgenic approaches.

Nontransgenic Male Sterility

Male sterility, the inability of a plant to produce fertile pollen, is a useful tool for hybrid breeding and hybrid seed production because self-pollination is prevented. Nontransgenic male sterility is used in sunflower, sorghum, and canola to produce hybrid crops; mechanical removal of pollen-bearing tassels is used more often to produce hybrid maize. Naturally occurring male sterility can be either "genic" or cytoplasmic. Genic male sterility results from mutations in nuclear genes. In most cases, cytoplasmic male sterility is based on mitochondrial genome rearrangements that lead to partial incompatibilities with the nuclear genome. Naturally occurring genic male sterility has been reported for many plant taxa. However, this trait is difficult to use in hybrid seed production because it is usually a dominant genetic trait and therefore is difficult to achieve without elaborate crossing schemes. Recently, genic male sterility has been produced by transgenic methods.

Experience with various crop plants demonstrates that male sterility is seldom perfect. Cytoplasmic systems, for example, can be overcome by nuclear restorer genes, temperature shifts, and other environmental factors (Burns et al., 1991; Hanna, 1989; Kumari and Mahadevappa, 1998; Michalik, 1978). Even though reversion to fertility occurs infrequently, seed producers must routinely patrol their plots to cull the occasional fertile plant. Thus, the opportunity for reversion is a disadvantage if pollen-mediated gene flow must be kept to a minimum. For some species, the low frequency of reversion could be manageable, however, and the use of cytoplasmic male sterility would improve the efficacy of physical containment (Pedersen et al., 2003). Patrolling and culling for revertants might not be appropriate for extensive, long-term field trials or for use with large plants such as trees that produce flowers at heights that are difficult to monitor. Genic male sterility might not be as susceptible to reversion. Reversions of nuclear male sterility genes would be expected only at the normal rate of background mutation. Thus, genic male sterility systems could be preferable to cytoplasmic systems if pollen-mediated gene flow must be kept to a minimum. There is substantial concern over transgene flow from GEOs to natural populations of related plants via pollen, so the use of male sterility is recommended whenever feasible.

Transgenic Male Sterility

Transgenic male sterility could allow for hybrid seed production to be introduced to crops for which natural genic or cytoplasmic systems do not exist. This could be a boon for productivity because hybrid seed crops often exhibit heterosis (hybrid vigor). Nuclear male sterility has been engineered in several species, including tobacco, rice, maize, alfalfa and *Brassica,* by using the *Bacillus amyloliquefaciens* barnase gene, which encodes a secreted ribonuclease that is cytotoxic. Zhan and colleagues (1996) fused the promoter of the rice-pollen-specific gene PS1 to the barnase gene. In transgenic tobacco plants, there was a range from reduced pollen fertility to complete sterility (Zhan et al., 1996).

One technical challenge of using cell ablation to obtain male sterility is that, if expression of the toxin is leaky—if it occurs in cells other than the flower buds—the plant can be damaged. If it is not possible to achieve sufficient specificity of expression to prevent secondary effects in non-targeted tissues, then ablation will not be a useful bioconfinement technique. Burgess and colleagues (2002) addressed the issue, and they showed that targeting specificity can be enhanced by engineering the barnase gene into 2 complementary fragments expressed from 2 promoters that overlap in expression. When coexpressed the 2 barnase fragments complement each other to reconstitute barnase activity (cytotoxicity). Male sterility resulted when expression of the partial barnase genes was targeted to the tapetum in a genetically engineered tomato. All 13 tomato progeny that inherited both transgenes were male-sterile. This dual-component system also allows genetically engineered lines to be used in hybrid seed production, because the progeny that inherit only a single barnase gene fragment are male-fertile. Crossing 2 lines homozygous for 1 barnase gene fragment each will produce a male-sterile hybrid.

Another approach to obtaining seed production in lines engineered for sterility is to introduce a "restorer" gene from a second line that can overcome the toxicity of the sterility gene. Restorer genes often are found in naturally occurring male-sterile plant populations. Jagannath and colleagues (2002) developed a transgenic line in Indian oilseed mustard (*Brassica juncea*) that was male-sterile by the action of the barnase gene but that was restored to fertility in the presence (expression) of a barstar gene, thus permitting both bioconfinement and the option for heterosis breeding.

Strengths

The biology of male sterility has been studied intensively by crop breeders. New transgenic methods could be more reliable than are other genetic mechanisms for inducing male sterility. Some nontransgenic methods also could be useful. When effective, male sterility can greatly reduce pollen-mediated crop-to-crop and crop-to-wild gene flow.

Weaknesses

Most types of male sterility are leaky, so it will be important to test the reliability of this trait in a representative range of environmental conditions. Also, transgenic methods could fail if gene silencing or recombination separates the confined gene from the sterility system. Another disadvantage is that male-sterile crops grown for seed will need sufficient incoming pollen to guarantee high seed set, and if transgenic male-sterile plants are pollinated by sexually compatible weeds, their progeny (if fertile) could establish weedy crop-wild populations that have undesirable transgenic traits. As with all methods of reducing pollen flow, the potential of seed dispersal and vegetative propagation should be examined to ensure adequate confinement.

Transgenes in Chloroplast DNA

A potentially powerful strategy to reduce or prevent the flow of transgenes through pollen grains of most flowering crops is to incorporate the transgenes into the plant chloroplast or plastid genome instead of incorporating them into the plant nuclear genome (Gray and Raybould, 1998; Maliga, 2001, 2002, 2003). In most flowering plants, chloroplast genes are maternally inherited and are not carried by pollen (Maliga, 2002). Plastid genetic transformation was first developed for tobacco (Svab et al., 1990), but the technology also has been used to transfer genes into *Arabidopsis* (Sikdar et al., 1998), rice (Khan and Maliga, 1999), tomato, (Ruf et al., 2001), and potato (Sidorov et al., 1999). Eventually, the technology could provide a useful bioconfinement measure in other species, including turfgrasses.

In addition to its use for bioconfinement, chloroplast gene transfer technology could offer commercial advantages over nuclear gene transfer methods. The production of a transgene product in chloroplasts is much higher than that for nuclear transgenes. Nuclear transgenes typically result in 2–3% total soluble proteins (reviewed by Kusnadi et al., 1997), whereas concentrations from chloroplast transgenes can be as high as 18% (Khan and Maliga, 1999). The production of chloroplast transgene products has been regarded to be 10- to 300-fold higher than that for genes transferred to nuclear genomes (Heifetz, 2000; Staub et al., 2000)—a production rate that could be especially useful for pharmaceuticals and industrial compounds. High concentrations of transgene-produced insecticides (*Bacillus thuringiensis* [*Bt*] toxins) could be needed for high-dose strategies to delay the evolution of insects that are resistant to plant-produced pesticides (Briggs, 1999). The greater production is possible because chloroplast transgenes are present as multiple gene copies per cell, and they are little affected by pre- or post-transcriptional gene silencing (Heifetz, 2000). A plastid genome could be transformed by homologous recombination, which allows the integration of transgenes at a specific site. That amount of precision could reduce the unintended phenotypic effects of transgenes, although it is not yet feasible for nuclear transformation.

The very high level of expression of a transgene or of stacking multiple transgenes in the chloroplast could disturb the function of normal plant physiology and therefore could hamper performance of the genetically engineered crop. Other limitations of chloroplast-based bioconfinement relate to questions about whether plastid DNA can be inherited paternally (*via* pollen). In gymnosperms, such as conifers, plastid genomes are transmitted primarily paternally; most flowering plants transfer plastid genomes maternally. However, approximately one-third of the flowering plants investigated (Mogensen and Rusche, 2000) exhibited some degree of paternal or biparental plastid inheritance. For example, rye (Mogensen and Rusche,

2000), chaparrel (Yang et al., 2000), kiwi (Tustolini and Cipriani, 1997), and the medicinal herb *Damiana* (Cipriani et al., 1995) are flowering plants with plastid genomes that are transmitted paternally. Some species exhibit polymorphism, with paternal inheritance accompanied by biparental and maternal variants (Rusche et al., 1995; Schumann and Hancock, 1991). In alfalfa, plastid genome inheritance is paternal (Keys et al., 1995) or biparental (Losoff et al., 1995; Zhu et al., 1993). In some flowering plants, the inheritance of plastid DNA can be even more complicated. For example, interspecific hybrids of calla lilies exhibited maternal chloroplast transmission in the first hybrid generation and either maternal or paternal inheritance in backcrosses (Yao and Cohen, 2000). Leakage can occur even in flowering plants for which paternal inheritance predominates (Avri and Edelman, 1991).

Chloroplast genes also can "jump" into the nuclear genome, although the chances of this happening seem remote. This has occurred over long periods of evolution and investigators are attempting to document gene exchange between cell plastids and the nucleus. A recent study documented a 0.0006 chance for such transfer, and such transplanted chloroplast genes were not expressed in the nucleus (Huang et al., 2003). Should the chloroplast-targeted gene construct contain chloroplast-specific transcriptional control systems, the targeted genes could integrate in the plant nuclear genome at a very low percentage, but they will not be functional and should not be a major concern (Daniell and Parkinson, 2003).

Despite the environmental and economic advantages, chloroplast transgene technology has not come into routine use, largely because the two gene transfer methods known for this technology have not been successfully applied to most crops. Biolistic bombardment of totipotent cells (cells that can produce whole fertile plants) and polyethylene-glycol-mediated naked DNA transfer into chloroplasts, followed by regeneration of whole selected transgenic plants, are still in development. More research is needed in this promising area.

Strengths

Chloroplast-specific transgenes would not be spread in the pollen of most cultivated plants. This approach could prevent transgene dispersal in pollen while preventing some of the disadvantages of male sterility, such as loss of pollen for cross-pollination.

Weaknesses

Technical difficulties have prevented this bioconfinement method from being feasible, and many types of desirable traits cannot be produced by proteins that are confined to chloroplasts. Also, the leakiness of the system will need to be demonstrated empirically on a case-by-case basis. Like all methods in this section, seed-mediated dispersal of the transgene is not prevented. In addition, wild-to-crop pollination could occur in some situations, resulting in hybrid progeny that are transgenic and potentially weedy.

Cleistogamy (Closed Flowers)

Cleistogamous flowers are those that never open; such flowers necessarily fertilize themselves (Lord, 1981). Therefore, creating plants with obligate cleistogamy has been mentioned as a possible bioconfinement method (e.g., Lu, 2003), although the committee is not aware of research on this topic. Theoretically, closed-flower varieties could be developed by selecting for sepals and other flower parts that encase the anthers and stigma. Obligate cleistogams would not be able to fertilize other plants, nor would they be able to be fertilized by other plants. With repeated self-pollination, obligate cleistogams that are derived from previous outcrossing would be subject to the genetic load uncovered by repeated inbreeding. Indeed, apparently no wild plant species produce flowers that are all cleistogamous; those species that produce them produce open "chasmogamous" flowers as well (Lord, 1981).

Strengths

At best use of obligate cleistogams would be an effective method of preventing gene escape by pollen.

Weaknesses

The method is not being developed, and perpetual self-fertilization could result in inbreeding depression. Transgene escape by seed or vegetative reproduction could still occur for plants with obligate cleistogamy.

Apomixis (Asexually Produced Seeds)

Apomictic plants reproduce asexually by clonally produced seed (Grant, 1981; Richards, 1997). Progeny produced by apomixis (agamopermy) are usually genetically identical to the parent, and therefore uniform within and between generations. Many plant species can reproduce sexually and asexually, but those that have dispensed with sex altogether are rare.

Some breeders and genetic engineers have sought to introduce apomixis to the final products of plant improvement to fix and propagate superior hybrid genotypes. Highly productive hybrid plants thus are produced easily, and there is no need to maintain inbred lines and cross them to create hybrids (Bock, 2002). Because obligately apomictic organisms do not require the fusion of female and male gametes to produce progeny and because they cannot be fertilized by gametes from another individual to create hybrid progeny, apomixis has been suggested as a bioconfinement method (Bock, 2002; Daniell, 2002; Gressel, 1999). If an organism is fully asexual and fully male-sterile, as for certain potato cultivars, then it cannot cross with other organisms.

Unfortunately, obligate apomixis is extremely rare; many apomictic plant species retain low to moderate sexual seed production (Grant, 1981; Richards, 1997). Furthermore, moderate to high pollen fertility is common in apomictic plants (Grant, 1981; Richards, 1997), and many apomictic species require pollination to stimulate seed formation, even though gamete fusion does not occur (this is called "pseudogamy" or "semigamy" [Richards, 1997]). If fertile pollen introduces an allele for apomixis into a natural sexual population, it could spread quickly through the natural population (van Dijk and van Damme, 2000). Sexual reproduction has short-term disadvantages that are attributable to the "cost of sex." A parent passes on 100% of its genes to asexually-produced progeny, but only 50% of their genes to outcrossed, sexually-produced progeny. Therefore apomyctic organisms have an automatic two-fold fitness advantage over sexual organisms. Population genetic models have shown repeatedly that apomictic organisms always replace outcrossing sexual organisms when all else is equal (e.g., Charlesworth, 1989; Marshall and Brown, 1981; Williams 1975; Maynard Smith, 1978).

If the apomictic allele is linked to the transgene, the resulting "selective sweep" could spread the transgene much more quickly and effectively than if the transgenic organisms were nonapomicts. The replacement of a sexual, genetically variable population with hybrid apomicts can lead to extinction by swamping (e.g., Ellstrand and Elam, 1993; Levin et al., 1996). This short-term advantage of asexuality is thought to account for the fact that if a species complex includes both sexual and apomict genotypes, the apomicts typically have a much wider distribution (e.g., Bierzychudek, 1985); for example, in the case of dandelions, the apomicts are widespread, and the sexual populations are mostly restricted to narrow refugia.

Strengths

Obligate apomixis with full male sterility will be an effective bioconfinement method only if the confinement goal is to prevent the formation of hybrid progeny.

Weaknesses

Studies of naturally occurring apomixis suggest that this bioconfinement method could be leaky. Given that even obligate apomicts still produce seeds, the method cannot be used for bioconfinement if the goal is to prevent the production of progeny that might disperse. Also, it will be important to confirm that apomictic GEOs cannot establish invasive populations.

Transgenes Absent from Seeds and Pollen

Nontransgenic Scions on Transgenic Rootstock

Many woody perennial crops, such as cultivars of grape, citrus, and avocado, are grown as grafted composites of two genotypes. The lower, root-bearing portion is the *rootstock*; the upper portion that bears flowers, fruits, and seeds is the *scion*. Lev-Yadun and Sederoff (2001) suggested that it is possible to graft nontransgenic scions of woody species onto transgenic rootstocks that have not yet reached reproductive age and that have had their branches pruned. Then, only nontransgenic reproductive structures are formed on those plants. In some species, vegetative propagules produced by the rootstock or other adventitious rootstock growth could still result in the production of transgenic flowers. Recognizing this, Lev-Yadun and Sederoff suggested double grafting those species that have adventitious growth from the roots, starting with a transgenic shoot grafted (an "interstock") to a wild-type rootstock and then grafting again with a nontransgenic scion. The result would be a transgenic section sandwiched between the nontransgenic material. This technique would be appropriate for trees tested and deployed on limited scales and those sold as grafts, such as fruit trees and ornamental trees.

Strengths

As long as the transgenic rootstock or interstock cannot produce branches that bear flowers or vegetative propagules—and as long as that can be tested and demonstrated in appropriate environments—this could prove to be a simple and effective method of bioconfinement.

Weaknesses

The technique could not be used for nonwoody species, and it can be applied only for transgenic traits that are expressed in the rootstock or interstock. This method would not be appropriate for large scale tests with forest trees.

Excision of Transgenes before Reproduction

Other researchers have proposed strategies for excising transgenes from plants before they begin sexual reproduction (Keenan and Stemmer, 2002) so a transgenic trait, such as herbicide resistance, could be expressed in the early stages of plant growth but the transgene would not spread in pollen or seeds. Unlike other V-GURTs, this method would not result in sterility, and growers could use saved seed (minus the transgene). A chemically induced or flower-specific promoter could be used to drive a recombinase enzyme that excised the transgene cassette for herbicide resistance; that transgene is located between two specific recombinase sites —*Cre* and *loxP* for example. This is an extension of other methods proposed for removing selectable marker transgenes (e.g., Zuo et al., 2001). However, it is much simpler to remove a marker gene in a few specimens before seed multiplication and commercial release than it is to remove genes from vast numbers of field-grown plants. Coaxing a promoter to work reliably in every flowering structure of every plant before pollen or seeds are formed is a major hurdle. Perhaps the strategy will be applicable in some crops at some time in the future, after technical problems with the method have been overcome. If the transgenic trait is not needed in the fruits or seeds, the approach could be useful.

Strengths

Transgene excision could be used to block the dispersal of transgenes in pollen and seeds, without requiring seed suicide.

Weaknesses

It could be extremely difficult to guarantee the reliability of the system. And it cannot be used for transgenic traits that must to be expressed in seeds or other reproductive structures.

Artificially Induced Transgene Expression

A promising method for reducing the effects of unwanted transgenes is to use a system in which the transgenic trait is activated by an artificial stimulus, such as a chemical spray (Figure 3-1; Daniell, 2002; FAO, 2002). Salicylic acid, for example, could be used to induce plants to produce pest-fighting compounds when pest populations reach a given threshold. Without the spray, the inducible promoter would not be activated and the plants would lack the trait for insect or disease resistance. Likewise, seed produced by the crop or plants that the crop crosses with would not express the trait, although they would carry the transgene construct. In another example, a chemical could be applied before the crop is sprayed with herbicide to induce herbicide resistance (Gressel, 2002). Because the trait expression is restricted by an inducible promoter, this approach is considered a T-GURT. The extent to which the system prevents the unauthorized acquisition of the trait will depend on the specificity of the stimulus, such as the type of chemical spray that is used.

Strengths

Use of artificially induced transgene expression could restrict transgene expression to limited periods. T-GURT constructs would not be expressed in other generations or in crop-wild hybrids unless a particular stimulus was applied. Thus, the spread of transgenes would not be prevented, but bioconfinement of particular traits would be possible.

Weaknesses

T-GURTs are still in the early stages of development and might not prove practical. Although they could give access to some transgenic traits that are useful on a transitory basis, other traits could require constant expression of the transgene to achieve a desired result (e.g., enhanced latex production in slow-growing guayule shrubs).

Reducing Gene Flow to Crop Relatives

Several approaches could be used to restrict the spread of transgenes to sexually compatible wild relatives and cultivars of a crop. None of those methods is in use, and they are not likely to achieve complete containment of transgenes. Nonetheless, they could come into use in the future, especially in combination with other biological and nonbiological confinement methods.

Repressible Seed Lethal Confinement

A group in Canada recently proposed a strategy for blocking gene flow to nontransgenic crops and wild relatives (Figures 3-1 and 3-2; Scherathaner et al., 2003). The approach involves inserting a "seed lethality" transgene into the crop plant's DNA. The transgene is tightly linked to a transgene that codes for a novel trait such as disease resistance. The plant is crossed with another plant that has a transgene that codes for repression of the seed-lethal gene. Then, selected F_1 offspring from the cross are used for seed multiplication and production for sale.

With this system, field-grown commercial plants produce viable seeds when they naturally self-pollinate or cross with each other in the field (Figure 3-2; 25% of the seeds are expected to be inviable because they would be homozygous for the seed-lethal transgene and would lack the repressor transgene). However, if a transgenic plant spontaneously crosses with a *different* crop variety, a wild conspecific, or a different species, all progeny with the seed-lethal–novel-trait construct will lack the repressor transgene, and all seed will be inviable. This occurs because the seed-lethal– novel-trait construct and the repressor transgene segregate independently. Ideally, the repressor transgene would be inserted at the same location on the homologous chromosome as the seed-lethal–novel-trait construct, so that the repressor and seed-lethal transgenes would not segregate together,

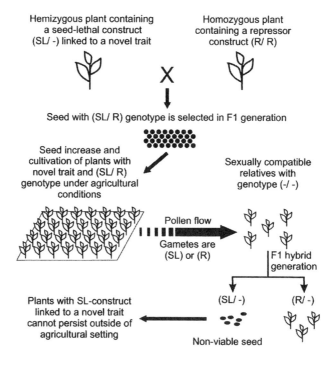

FIGURE 3-2 Repressible seed–lethal bioconfinement. Adapted from Schernthaner et al., 2003. SL, seed-lethal gene; R, repressor gene.

except as a result of rare recombination events (chromosome crossovers). During meiosis and gamete formation, the haploid pollen and ovules inherit one transgene construct or the other, but not both. Thus, outgoing crop pollen and incoming noncrop pollen produce inviable seeds with this system. Presumably, all seeds would complete normal development up to the point of maturation, and germination would be the only step affected by the seed-lethal transgene.

The repressible seed-lethal method is similar to V-GURTs, such as the terminator technology, but it differs in that farmers can obtain viable seed from a transgenic crop. Farmers in developing countries would be able to save and distribute seeds for future use, although as much as 25% of a seed crop could be inviable, which is unlikely to be acceptable. Farmers would not be able to cross a transgenic crop with local varieties, so in this sense their use of agronomically valuable crop genes is still restricted, as it is with other GURTs.

Strengths

The extent of pollen-mediated transgene dispersal would be greatly reduced or eliminated because offspring that inherit both the new transgenic trait and the tightly linked seed-lethal transgene would not inherit the repressor transgene and thus would not pass the novel transgene to their offspring. At the same time, unauthorized use of the transgenic crop for further breeding would be difficult or impossible, which is an advantage.

Weaknesses

This method is still in the early stages of development, and several technical hurdles must be overcome before it can be used as a bioconfinement method. For example, site-specific insertion of transgenes has yet to be achieved in plants. Partial confinement is possible, though, as long as the transgenic constructs are located on homologous chromosomes. Concerns about the consequences for nearby relatives of producing dead seeds from the crop are similar to those for terminator and related transgenic sterility methods. Continued use of this system in a single locale could lead to the introgression of the repressor gene into nearby natural populations. When it reaches a high enough frequency, the presence of that allele would render the method ineffective in preventing introgression into that population. Finally, the method does not prevent seed-mediated dispersal of the transgene, for example by natural seed dispersal, spillage during harvest and transport, seed mixing, or international distribution of food aid.

Cross-Incompatibility

Crosses between species, both plant and animal, often fail because of events that prevent fertilization or embryo development (Futuyma, 1998). Such barriers to hybridization are called "cross-incompatibility." Crosses between cross-incompatible species either fail entirely or fail most of the time. In plants, cross-incompatibility can be expressed as the inability of pollen to germinate, as abnormal pollen tube growth, as failure of pollen tubes to penetrate the ovary, or as the spontaneous abortion of seeds and fruits after fertilization (Levin, 1978). It has been suggested that alleles for cross-incompatibility could be identified and moved into transgenic varieties to prevent them from mating with other varieties (Evans and Kermicle, 2001). Research on "lock and key" approaches to prevent pollen from fertilizing ovules on other varieties of sexually compatible plants is still very preliminary, as described at a recent workshop (Arcand, 2003).

Strengths

If absolute, bilateral incompatibility is created between different lineages, species, subspecies, and so on, natural hybridization can be prevented.

Weaknesses

Such bioconfinement methods have not yet been created and tested. Because some cross-incompatibility barriers can be breached by environmental factors, such as high temperature or the presence of pollen from a compatible relative (e.g., Richards, 1997), it is possible that engineered incompatibilities might also be environmentally labile. Obviously, cross-incompatibility will not prevent the movement of transgenes in seed or vegetative propagules created by the transgenic plant.

Chromosomal Location in Allopolyploids

This technique involves placing transgenes in chromosomes that would be preferentially excluded in crop-wild progeny because of problems with chromosomal pairing at meiosis. Many crops—bread wheat, peanuts, and coffee—are allopolyploids that house multiple genomes derived from different sources. For example, bread wheat is a hexapolyploid with three paired sets of homologous chromosomes (2n = 6x = AABBDD). Because there is an even number of matching sets of chromosomes, bread wheat plants undergo normal meiosis and gamete formation—as though they were diploids. Often, only one genome of the crop is homologous with that of a related weed. Alleles on those chromosomes will be transmitted to the weed by backcrossing with hybrids. But it has been thought that alleles on the other genomes would face a considerable barrier to transmission. For example, bread wheat, *Triticum aestivum*, and jointed goatgrass, *Aegilops cylindrical*, which is a weed, share the same D set of chromosomes. Bread wheat's A and B sets are not homologous with the chromosomes of jointed goatgrass. If a novel allele is incorporated into either the A or the B chromosomes of bread wheat, they should be preferentially excluded over the generations of backcrossing to the wild parent and have considerable difficulty introgressing into wild populations (Gressel, 1999). In fact, Lin (2001) noted that a transgene that confers herbicide resistance was inherited by jointed goatgrass when it occurred on the D genome, which is shared by both taxa, but not when it occurred on the unshared B genome. This is encouraging, but it will still be important to measure the frequency of rare episodes of recombination that could allow transgenes to move into non-homologous sets of chromosomes. Apparently, other experimental work has revealed that chromosomal location is not a sure safeguard in allo-polyploids of wheat (Zemetra, unpublished data).

Some data are available for a similar cytogenetic situation involving another allopolyploid crop. Oilseed rape, *Brassica napus* (2n = 4x = AACC), shares one set of chromosomes (the A set) with the weed *B. campestris* (= *B. rapa*) (2n = AA). Therefore, it might be expected that if alleles of concern were placed on rape's other set of chromosomes (the C set) they would be preferentially excluded in the wild (Gressel, 1999). When Metz and colleagues (1997) observed a strong decrease in the frequency of a transgene in progeny resulting from a backcross of a *B. napus* × *B. campestris* hybrid to *B. campestris*, they explained that decrease as the result of such preferential exclusion. Tomiuk and colleagues (2000) examined the situation further and reached a different conclusion. First, they found that other cytogenetic data (Fantes and Mackay, 1978) showed no preferential exclusion of the C genome in backcrosses. Second, they created a model to examine the data of Metz and colleagues (1997) more closely. They found that an alternative, equally parsimonious, hypothesis could not be excluded: that the "decrease in the frequency of transgenic plants within the first backcross generation can also easily be explained by selection against transgenic A-chromosomes of *B. napus*" (Tomiuk et al., 2000). They conclude, "without more detailed genetic information...no decision can be made in favor of the A- or C-genome as the safer candidate with respect to the introgression of transgenes into wild populations" (Tomiuk et al., 2000). Clearly, there is a need for more experimental work of this type. As with some of the other proposed bioconfinement methods, this one would work primarily for preventing or reducing introgression from the transgenic organism to wild populations.

Strengths

If this method were fully effective, it would prevent backcrossing to the wild parent, limiting introgression of transgenes into populations of non-transgenic wild relatives.

Weaknesses

The technique would not necessarily limit transmission of transgenes into the F_2 progeny of crop-wild hybrids.

Fitness Reduction in Transgenic Crop-Wild Progeny

Gressel (1999) proposed "tandem constructs" for "transgenic mitigation (TM)" in crops. His idea is to link to both sides of a transgene alleles that would confer a substantial disadvantage to a weed or volunteer. Here is the overall scheme:

> (1) Tandem constructs of genes genetically act as tightly-linked genes, and their segregation from each other is exceedingly rare, (2) There are traits that are either neutral or positive for a crop that would be deleterious to a typical or volunteer weed, or to a wild species; and (3) Because weeds are strongly competitive amongst themselves and have large seed outputs, individuals bearing even mildly harmful traits are quickly eliminated from populations. Even if one of the TM genes mutates, is deleted, or crosses over, the other flanking TM gene will remain providing mitigation (Gressel, 2002).

Traits that would be beneficial under cultivation, but detrimental to plants in the wild, are common in modern agronomic crops: lack of secondary seed dormancy, uniform ripening, lack of shattering, dwarfing, or susceptibility to a specific herbicide (Gressel, 1999, 2002). For tandem constructs to be effective, the traits conferred by the flanking alleles must be dominant relative to their counterparts in wild plants. Although the genetic basis and chromosomal location for such traits are unknown for most crops, data are beginning to accumulate (e.g., Burke et al., 2002; Gepts, 2001; Gressel, 2002). To illustrate how a tandem construct might work, Al-Ahmad and Gressel (2002) created an experimental model system. They transformed tobacco with an herbicide resistance gene linked to a dwarfing gene. The dwarfed plants proved competitively inferior to the wild-type segregants, and only at the lowest density treatment did the dwarfs form flowers.

Strengths

If effective in a wide variety of environments, TM would limit the introgression of transgenes in the wild. Although the tandem-construct concept was designed specifically to prevent crop alleles from establishing in free-living populations, it is clear that it can be extended to many bioconfinement situations as long as the transgenic mitigator confers a substantial disadvantage to organisms that bear it in an environment for which the primary transgene was not intended.

Weaknesses

This method depends on using linked alleles that are harmful to wild relatives but neutral or beneficial for crops . Although they will frustrate the evolution of weediness, if introgression of the biocontrol alleles into a very small population did occur by pollen or seed swamping, then depressed fitness of future generations of that population could result, increasing the risk of extirpation of that population, which would be a concern if an endangered wild relative is involved.

Phenotypic and Fitness Handicaps

Many cultivated plants are highly domesticated, requiring special conditions, such as amended soil; irrigation; and protection from weeds, pests, and pathogens, to survive and reproduce. Maize, soybean, tomato, and many other food crops fall into this category because they rarely, if ever, become naturalized. In contrast, other species (Bermuda grass, raspberry, poplar, spruce) are more similar to their wild progenitors and are therefore more likely to establish feral populations. Conventional and transgenic methods can be used to select cultivated genotypes in both groups of plants that are increasingly "domesticated," in the sense that they need specific human intervention to be able to survive and reproduce. Handicap methods could be most useful in species that are relatively undomesticated, but even species like maize could become easier to confine by adding handicaps that restrict growth under standard agricultural conditions. Handicap strategies are similar to the tandem construct method for lowering the fitness of any feral or wild plants that carry unwanted transgenes, but they are more general in that the crop plants have new features that make them more dependent on specialized growing conditions.

One of many possible approaches to establishing a biological handicap is to select for plants that are "chemically dependent." Auxotrophs are a class of mutants that depend on the exogenous supply of a nutrient that arises from a mutation in a biosynthetic pathway. Such fitness-reducing mutations could be selected after mutagenesis or created by transformation. Numerous examples of plant auxotrophs created by exposure to mutagens have been reported in the literature (Blonstein et al., 1988; El Malki and Jacobs, 2001; Fracheboud and King, 1988, 1991; Meinke, 1991; Wright et al., 1991). Few genetically engineered auxotrophs have been reported, but they are likely to become plentiful as more developmentally important genes are identified. One possible strategy is to use knock-out mutants that delete a vital biological process, such that the plant requires artificial growing conditions for survival. In another approach, Baroux and colleagues (2001) reported developmental mutations in *Arabidopsis* that were created by transactivation of barnase and that affected the embryos and growing shoots of adult plants. When the barnase protein was expressed during embryogenesis it destroyed most of the cellular RNA, leading to the production of sterile seed. To maintain genetically engineered lines that carry auxotrophic mutations (for survival during field tests and production), it also will be necessary to develop transgenic methods to overcome the mutation. Thus, the creation of or inclusion of auxotrophic or developmental mutations could be one way to prevent transgene escape into the environment.

Another possibility would be to select for life history traits and morphological variants of fast-growing trees, so they have greatly restricted branching and short stature at maturity (Mann and Plummer, 2002). If miniature poplar trees were selected for rapid growth and commercially important traits, such as low lignin content, those genotypes and their progeny might be unfit for survival other than in intensively managed settings. By adding a back-up confinement method, such as male sterility, gene flow and persistence of genetically engineered traits in other poplar populations could be so low as to become negligible.

Strengths

Auxotrophy and other handicap strategies might contribute eventually to integrated confinement methods.

Weaknesses

The methods described above are still in the early stages of development. To be effective, it will be important to ensure that bioconfined transgenes remain tightly linked to handicap traits and that they do not segregate with weedier traits after episodes of sexual reproduction and gene flow. Although they will frustrate the evolution of weediness, if introgression of the biocontrol alleles into a very small population did occur by pollen or seed swamping, then depressed fitness of future generations of that population could result, increasing the risk of extirpation of that population, which would be a concern if an endangered wild relative is involved. As with all types of bioconfinement, the leakiness of the methods should be determined empirically under realistic field conditions before they are used to prevent the spread of transgenes.

Reducing Exposure to Transgenic Traits

In some cases, the reason for choosing bioconfinement of a transgene could be to reduce human or environmental exposure to transgene products. This sometimes can be accomplished using special promoters, such that the transgene is expressed in some parts of a plant but not in others. Plant tissue and organ-specific gene expression can be used to produce a heterologous protein (a protein conferred by a transgene) that occurs only or mainly in specific tissues or organs. Most transgenic crops have constitutive promoters that allow the transgene to be expressed at all times throughout the plant. In the future, many more options will be available, including chemically induced, tissue-specific promoters (e.g., Mett et al., 1996). Here, we review a few examples from the large body of research findings on tissue-specific promoters. Some of them also could be useful in bioconfinement methods, such as inducible lethality in seeds, which involve targeted blocking of plant growth and development.

Green-Specific (Chloroplast-Targeting) Gene Expression

The photosynthesis-specific promoter of the ribulose 1-5, bisphosphate carboxylase (rubisco) gene of tomato has been used to express a *gus* gene (a marker for transgene expression) in green tissues of apple trees (Gittins et al., 2000). That promoter also has been used to regulate other genes in *Arabidopsis* and maize (Poirier et al., 1992; Zhong et al., 2003). Building on this research, it should be possible to keep transgenic seeds, pollen, and roots free from specific transgene products. Because pollen does not contain chloroplasts, photosynthesis-targeted gene expression could be an ideal method for reducing exposure to transgenic products in pollen. A better but more difficult method for achieving green-specific gene expression is to transfer genes directly into the chloroplast genome rather than to the nuclear genome. (See above section on chloroplast transformation.)

Roots and Tuber-Specific Gene Expression

Promoters specific to roots have been used to produce heterologous proteins that are not produced in other parts of the plant (Sakuta and Satoh, 2000; Yamamoto et al., 1991). The roots of carrots and a few other species contain carotenoids, so the genes of interest could be expressed using a carotenoid-specific gene promoter at the same time the presence of the heterologous gene products is minimized in other parts of the plant (Fraser et al., 1994). Tuber-specific promoters have been used in potatoes and other crops. For example, human interleukin genes under the control of a patatin promoter have been expressed in vitro in potato microtubers. In that experiment, the microtubers functioned as bioreactors to produce large amounts of interleukin (Park and Cheong, 2002). Potato tubers also have been used as bioreactors for production of a sucroselike compound using a gene from *Erwinia rhapontici* (Boernke et al., 2002).

Another root-specific location for transgene expression in legumes is the root nodule, within which symbiotic bacteria fix nitrogen. A complete soybean leghemoglobin gene was exclusively expressed in root nodules of transgenic *Lotus corniculatus*. In this case, gene expression was observed after transgenic roots were infected with a nitrogen-fixing bacterium (Stougaard et al., 1987).

Vascular-Tissue-Specific Gene Expression

Some insects, such as aphids and hoppers, that suck plant sap by feeding on plant vascular tissues, could be controlled by biopesticides that are expressed only in the plant vascular systems. Several strategies for vascular-tissue-specific gene expression have been identified in experimental systems. When a marker gene (*gus*) was expressed with the maize streak virus coat protein promoter in transgenic rice, the GUS protein was produced only in the vascular tissues, particularly in phloem-associated tissues (Mazithulela et al., 2000). Also, when the *Commelina* yellow mottle virus promoter was used to express the *gus* gene in transgenic oat, vascular-specific production of the GUS protein was observed in shoots, leaves, floral bracts, roots, and vegetative parts of ovaries but not in reproductive cells (Tolbert et al., 1998). Phloem-specific gene expression also was produced in transgenic rice plants using the RTBV (rice tungro bacilliform virus) promoter (Yin et al., 1997).

Flower- and Fruit-Specific Gene Expression

To control insects and pathogens that attack young flowers, transgenes can be expressed in the sepals and not in anthers, seeds, or other plant parts. For example, experiments on *Forsythia X intermedia* cv. Spring Glory using the *ans* gene showed that *ans* is exclusively expressed in sepals at the early stages of flower development (Rosati et al., 1999). A similar method of tissue specificity of gene expression was used to express the genes of interest in carotenoid-rich parts of plants, such as tomato fruits, while avoiding the presence of the heterologous gene products in other plant parts (Fraser et al., 1994).

Pollen-Specific Gene Expression

The need for pollen-specific transgene expression may be relatively uncommon, unless the transgenic trait is needed specifically in pollen. In some applications, the desired trait might be reduced allergenicity. For example, antisense technology can be used to reduce or eliminate the harmful expression of a naturally occurring gene in pollen. The allergic asthma effect of ryegrass pollen was reduced by inserting a pollen-specific promoter to drive an antisense gene that silenced an allergen gene (Bahalla et al., 1999).

Seed-Specific Gene Expression

Seed-specific gene expression can be used to produce the gene product only in the seed parts such as the embryo or the aleuron. The method could be useful for transgenes that confer improved seed quality or protection from insects during seed storage. Two barley aleuron-specific promoters from genes that encode lipid transfer protein (Ltp1) and chitinase (Chi26) were used to express the *gus* marker gene in grains of transgenic rice (Hwang et al., 2001). Similar experiments to demonstrate the efficacy of other seed-specific promoters have been carried out in soybean, tobacco, and bean (Baeumlein et al., 1987; Cho et al., 1999; Ellis et al., 1988; Iida et al., 1995). Late-acting, seed-specific promoters also can be used to kill the seeds just before they are fully ripe, as in the terminator and related applications described above.

Strengths

Tissue- and organ-specific promoters could be useful for reducing the amount of novel protein that a plant produces or for targeting specific organs, such as anthers, in order to interfere with their development.

Weaknesses

Some of the currently available tissue- and organ-specific promoters are not as precise or effective as would be required to avoid transgene expression in other parts of the plant. Basic research is still needed on the regulation of gene expression, including studies of genes that could be used for tissue- and organ-specific gene expression and for genes that could be turned on or off by chemical intervention or by other methods, such as exposure to extreme temperatures. In many examples described above, bioconfinement would not be improved, although exposure to unwanted transgene products could be reduced.

Choice of Alternative Organisms or "Abstinence"

In many cases confinement can be obtained from specific aspects of the biology of an organism. Some GEOs, such as vaccine-producing microalgae, can be grown and harvested indoors, thereby obviating the need to develop bioconfinement for field conditions. If field releases are essential for a given genetically engineered application, some characteristics of the transgene "host" species, such as whether it is traded as a commodity crop, like corn and soybean, could greatly influence whether bioconfinement is needed. Likewise, if all possible plant species that host a given genetically engineered application require strict containment rather than confinement, plans to produce field-released plants should be abandoned. Choosing which GEOs to develop and which to abandon is effectively a form of bioconfinement.

The biosafety reasons for choosing a particular organism and the place of its deployment—in the field or indoors, for example—are varied. The choice of an appropriate plant for producing an industrial compound must consider whether it could cause harm to humans if consumed. A plant that typically is grown to produce a common food product would be a poor choice for engineering to produce that compound, unless the plant were to be grown under stringent conditions of confinement (Ellstrand, 2003b). This is an important issue for any novel compound or GEO for which zero tolerance is given for bioconfinement failure.

Increased security in bioconfinement can be obtained in three ways that involve the choice of the system: choosing a different organism to engineer, choosing not to grow genetically engineered plants outdoors, or choosing to abandon the project. Each is examined below, using the example of a novel industrial compound that must not enter the human food supply:

Organism choice. Choosing an organism that is not used for food or feed could prevent that compound from entering the human food chain. Many nonfood plants have been successfully transformed, including tobacco, petunia, and duckweed. Likewise, plants, such as belladonna, are known to be toxic, and they can be used because they already are avoided as a food source.

Field release choice. In many cases, valuable industrial compounds can be grown in high concentrations in plants. It is much easier to monitor plants that are grown and processed indoors and to control their reproductive processes. Likewise, alternative organisms that can be grown indoors, such as microbes, could be used to produce some compounds. For example, genetically engineered insect larvae could be grown in vats under strict biosafety procedures and processed before reaching adulthood.

Choice not to proceed. Growing organisms that produce extraordinary amounts of a toxic compound might require such stringent bioconfinement that a project would not be cost-effective. Considered broadly, the decision not to develop a given GEO is a form of bioconfinement. The committee recognizes that some biotechnology companies already have decided not to proceed with projects because of intractable biosafety issues. Also, some reasons for applying genetic engineering in the first place can be addressed using alternative approaches, such as improving integrated pest management, obviating the need to develop genetically engineered pest-protected plants.

Strengths

The careful choice of which organism to develop, whether to proceed with field release, or whether to abandon an idea or project altogether constitute bioconfinement. By making the decision early, expensive and difficult confinement options are rendered unnecessary.

Weaknesses

Choosing a new organism can set a project back in time and cost, and there is always the chance that the organism will not prove commercially viable. Baseline information and optimal breeding and cultivation techniques for the new organism might need to be developed. The choice of not growing a GEO outdoors can limit profitability, especially if techniques for indoor cultivation must be developed. Choosing not to proceed with a project is an even more difficult decision economically, especially if there has been substantial early investment. Moreover, abandonment of a project could prevent some benefits from being realized.

The following sections include a discussion of bioconfinement options for genetically engineered trees, and short overviews of related topics in grasses and algae. Trees and grasses have reached the field-testing stage of development. Seaweed and other macroalgae are just beginning to be investigated and produced. Some features of those organisms pose unique challenges for effective bioconfinement, whereas other issues are common to many types of GEOs. In addition, although some species have a long history of cultivation and genetic improvement, others are essentially undomesticated, so there is little baseline knowledge of relevant biological information.

GENETICALLY ENGINEERED TREES

Two fundamentally different technologies are used to transfer genes into trees: *Agrobacterium*-mediated gene transfer and microprojectile bombardment (or biolistics). The naturally evolved plant transformation system of *Agrobacterium* is considered more reliable for producing stable transformants, so it is the system of choice for most species. However, not all trees are susceptible to *Agrobacterium*. Conifers, for example, are especially resistant, so biolistics has been the method of choice (Klein et al., 1988). An advantage of biolistics is the relative ease of cotransforming genes on separate plasmid vector DNA (Bishop-Hurley et al., 2001), and cotransformation lends itself well to multiple-component systems for bioconfinement that prevent dispersal of transgenes (Table 3-2).

Although particle bombardment is easy, the approach tends to deliver several transgene copies into each recipient cell, and they often integrate as tandem repeats. This arrangement frequently leads to gene silencing or excision (loss) of the transgene (see Box 3-1). *Agrobacterium*-mediated transformation is more likely to result in the integration of a stable, single, full-length copy of the transgene in a region of the chromosome where genes are actively expressed. *Agrobacterium*-mediated gene transfer is effective for use in many fruit and nut trees (reviewed by Trifonova and Atanassov, 1996) and in hardwood timber species. Conifer species are beginning to yield to *Agrobacterium*-mediated gene transfer as well, through the use of multiple copies of virulence genes (Wenck et al., 1999).

Bioconfinement of Trees

In many cases, confinement of genetically engineered trees is not necessary. In fact, the transfer of resistance genes could be used to restore populations of trees that are threatened by exotic pests and insects (Adams et al., 2002). Examples include American chestnut and American elm, which were lost in the past century; Fraser fir and eastern hemlock, which have declined recently across much of their native ranges in the eastern United States; and oak trees, which are experiencing sudden death in western states. Genetic engineering could help restore those species within a manageable period and without the genetic dilution that occurs with sexual hybridization. If the introduction of genetically engineered trees is restricted to areas they once inhabited, growth and eventual spread should remain restricted to their natural ranges. Thus, beyond limiting spread to the sites of introduction, confinement techniques might not be necessary in restoration projects.

Another trait that might not always require confinement is lignin modification. Lignin is important to the structural integrity and adaptive strategies of vascular plants, but it is problematic for agroindustrial use of crops

TABLE 3-2 Genetically Engineered Woody Plants, Permits Approved by APHIS for Field Tests in the United States, 1989–2003

Organism	Phenotype (Number of Submissions)	Gene
Apple	AP - Flowering time altered (1)	LFY
	BR - Fire blight resistant (7)	Cec-B or Att-E
	FR - Apple scab resistant (3)	CHT
	IR - Oblique banded leafroller (1)	CHT; CHI
	IR - Coleopteran resistant (1)	CryIA(b) and CryIA(c)
	IR - Lepidopteran resistant (7)	CryIA(c)
	PQ - Brown spot resistant (2)	PPO
	PQ - Ethylene synthesis reduced (1)	ACCS antisense
	PQ - Fruit ripening altered (4)	SAMT or ES; ACCS antisense
	PQ - Sugar alcohol levels increased (5)	SPDH or SDH
Avocado	FR - Fungus resistance (1)	Def
Citrus sinensis X *Poncirus trifoliate*	BR – *Xanthomonas campestris* resistant (1)	LYZ
Coffee	PQ - Caffeine concentration reduced (1)	XMT antisense
	PQ - Ethylene production reduced (2)	ACO or ACS
Cranberry	IR - Lepidopteran resistant (1)	CryIA(a)
Eucalyptus grandis	HT; MG (1)	CBI; GUS
Grape	BR - crown gall resistant (4)	CBI
	FR - Botrytis resistant (3)	CBI
	IR - Lepidopteran, *Criconemella*, *Meloidogyne* (1)	CryIA(c); GNA
	FR - Powdery mildew resistant (10)	PGUS; LGB; PGLC; CHT; PGL
	MG, SM (4)	AHAS variant or ALS; CBI, NPTII
	PQ - Improved fruit quality (1)	ALS; CBI
	VR - Closterovirus resistant (3)	CBI
	VR - Nepovirus resistant (2)	CBI
	VR - Nepovirus resistant; B - Closterovirus resistant (1)	CBI
	VR - CBI (2)	CBI
Grapefruit	BR - Citrus canker resistant (1)	SBP
	IR - Aphid resistant (1)	GNA
	MG (1)	GUS; NptII
	VR - Closterovirus resistant (2)	CTV-CP

continued

TABLE 3-2 Continued

Organism	Phenotype (Number of Submissions)	Gene
Papaya	FR - Fruit rot, powdery mildew, *Phytophthora* (1)	CHT
	IR - Leafhopper resistant (1)	GNA
	PQ - Ethylene production reduced (2)	ACSp; Cre
	VR – PRSV resistant (15)	PRSV-CP
Pear	AP (1)	Rol
	BR - Fire blight resistant (1)	Cec-B
	PQ - Fruit ripening altered (3)	SAMT
Persimmon	AP - Drought and cold tolerant (1)	COX; SORS; GUS
	FR (1)	PGIP; GUS
	IR - Lepidopteran resistant (1)	CryIA(c)
	MG (1)	GUS
Pine	MG (17)	GUS; NptII; CBI
	PQ - Decreased lignin (1)	CBI
Plum	PQ - Ethylene production reduced (1)	ACOp
	VR - PPV resistant (2)	PPV-CP
Poplar	AP - Altered lignin biosynthesis (4)	4CL, OMT, C4H, COMT
	BR - Crown gall resistant (1)	IAAm
	FR - *Septoria* and others (2)	PGL; OXA
	FR - General (*Venturia*, etc.) (2)	BAC
	HT - CBI (8)	CBI
	HT - Glyphosate tolerant (19)	CBI; EPSPS; or GOX
	HT - Glyphosate; phosphinothricin; PQ (1)	Barnase; Barstar; PAT MS1; CBI
	HT – Phosphinothricin; MG (1)	PAT
	IR - Coleopteran resistant (13)	CBI or CryIIIA
	IR - Lepidopteran resistant (1)	CryIA(c)
	IR - Leaf beetle resistant (1)	CryIIIA
	MG and SM (1)	NptII; CAT
	MG only (4)	GUS; CBI
	OO - Cell wall altered (1)	CBD
	OO - Flowering time altered (1)	CBI
	OO - Sterility (3)	Barnase; DTA; LFY
	PR (2)	P450; MIR
	PN (1)	GS
Populus deltoides	HT (7)	CBI
	MG (2)	GUS; NptII
Rhododendron	FR - *Phytophthora* resistant (2)	Magainin
	MG (1)	GFP; NptII

TABLE 3-2 Continued

Organism	Phenotype (Number of Submissions)	Gene
Raspberry	FR (1)	PGIP
	FR - Fruit rot resistant (1)	PGIP
	VR - RBDV resistant (3)	RBDV-MP
	VR - ToRSV resistant (1)	ToRSV-CP
	PQ - Fruit ripening altered (3)	PGIP; PGIP and SAMH
Service berry	IR - Lepidopteran resistant (1)	CryIA(c)
Spruce	IR - Lepidopteran resistant (1)	CryIA(c)
Sweetgum	AP - Altered plant development (1)	CBI
	AP - Fertility altered (2)	CBI
	HT (4)	CBI
	HT - 2,4-D tolerant (2)	Tfd
	HT - Glyphosate tolerant (3)	CBI
	HT - Phosphinothricin (1)	CBI
	MG only (4)	GUS
Walnut	AP - Adventitious root formation (2)	rol and CBI
	AP - Cutting rootability increased (1)	Rol
	AP - Flowering altered (1)	LFY
	BR - Bacterial leaf blight resistant (1)	TMK
	FR; IR; VR (1)	LRV-CP; LEC; SAR; rol
	IR – Lepidopteran resistant (5)	CryIA or CryIA(c)
	NR - *Pratylenchus vulnus* resistant (1)	GNA

NOTE: Field test data downloaded from Information Systems for Biotechnology, http://www.nbiap.vt.edu/cfdocs/fieldtests2.cfm Feb. 19, 2003; updated May 23, 2003; does not include submissions denied or withdrawn.

Phenotype Key: AP, agronomic properties; BR, bacterial resistance; FR, fungal resistance; HT, herbicide tolerant; IR, insect resistant; MG, marker gene; NR, nematode resistance; OO, other; PN, plant nutrition; PR, bioremediation; PQ, product quality; SM, selectable marker; VR, virus resistance.

Gene Key: 4CL, 4-Coumarate:CoA ligase antisense gene from poplar; ACO, ACC oxidase antisense from coffee; ACOp, ACC oxidase antisense from *Prunus*; ACS, ACC synthase antisense from coffee; ACSp, ACC synthase antisense from papaya; AHAS, acetohydroxyacid synthase; ALS, acetolactate synthase; Att-E, attacin gene from *Hyalophora cecropia*; BAC, bacteropsin gene from *Halobacterium halobium*; BARNASE, barnase gene; barstar, barstar gene from *Bacillus amyloliquefaciens*; C4H, Cinnamate 4-hydroxylase gene from *Populus tremuloides*; CAT, chloramphenicol acetyltransferase gene from *E. coli*; CBD, cellulose binding protein gene from *Clostridium cellulovorans*; CBI, confidential business information; Cec-B, cecropin gene from *Hyalophora cecropia*; CHI, chitobiosidase probably of fungal origin; CHT, chitinase probably of fungal origin; CLRV-CP, coat protein gene from CLRV; COMT, caffeate O-methyltransferase gene from *Populus tremuloides*; COX, choline oxidase; Cre, recombinase from Bacteriophage P1; CryIA(c), CryIA(c) crystal toxin gene from Btk; CryIIIA, CryIIIA crystal toxin gene from Bt; CrylA, crystal toxin gene A from *Bt*; CTV-CP, coat protein gene from CTV; Def, defensin from *Arabidopsis thaliana*; DTA, diptheria toxin

continued

TABLE 3-2 Continued

A gene from *Corynebacterium diptheriae*; EPSPS, 5-enolpyruvylshikimate-3-phosphate synthase; ES, ethylene forming enzyme from apple; GFP, green fluorescent protein from *Aequorea Victoria*; GNA, lectin gene from snowdrop (*Galanthus nivalis* agglutinin); GOX, glyphosate oxidoreductase gene; GS, glutamine synthase gene; GUS: *E. coli* β-glucuronidase gene; HYR, hygromycin phosphotransferase gene; IAAm, IAA monooxygenase gene; LEC, lectin genes from barley, rubber tree, and/or stinging nettle; LFY, leafy homeotic regulatory gene from *Arabidopsis thaliana*; LGB, lignan biosynthesis protein gene from pea; LYZ, lysozyme gene from cow; magainin, magainin gene from *Xeanopus laevis*; MIR, mercuric ion reductase from *E. coli* ; MS1, male sterility protein gene from *Populus trichocarpa*; NPTII, neomycin phosphotransferase gene from *E. coli*; OMT, O-methyltransferase gene from *Populus tremuloides*; OXA, oxalate oxidase gene from *E. coli*; P450, cytochrome P450 gene from man; PAT, phosphinothricin acetyl transferase gene from *Strep. hygroscopicus*; PGIP, polygalacturonase inhibitor protein from bean; PGL, anitmicrobial peptide gene from wheat; PGUS, β-glucuronidase from pea; PGLC, B-1,3-glucanase antisense from pea; PPO, Polyphenol oxidase from Apple; PPV-CP, Coat protein gene from PPV; PRSV-CP, Coat protein gene from PRSV; RBDV-MP, nonfunctional RBDV movement protein; rol, rol hormone gene from *Agrobacterium rhizogenes*; SAMH, S-adenosylmethione hydrolase from *E. coli*; SAMT, S-adenosylmethionine transferase from E. coli; SAR, systemic acquired resistance gene from tobacco; SBP, synthetic binding peptide to *Xanthomonas*; SORS, sorbitol synthase from apple; SDH, sorbitol dehydrogenase from apple; SPDH, sorbitol 6-phosphodehydrogenase from apple; Tfd, monooxygenase gene *from Alcaligenes eutrophus*; TMK, receptor kinase gene from rice; ToRSV-CP, ToRSV coat protein gene; XMT, xanthosine-N7-methyltransferase antisense from coffee.

BOX 3-1
Stability of Transgenic Confinement

Stable gene expression is a necessity for bioconfinement that is based on transgenic approaches. Expression must be stable throughout the lifespan of the organism, and it must be adequate to accomplish confinement. For inducible or regulated genes, there must be confidence that expression will reach needed levels at the appropriate times, year after year for perennials. Just as transformants are selected for strong, stable expression of agronomic transgenes during crop development, so too should sufficient evaluation be given to the stability of the engineered bioconfinement method. In the context of using transgenic methods for bioconfinement, instability that is not detected before field releases is clearly undesirable. The question of stable integration and expression of foreign genes is important for long-lived species, such as trees and turfgrasses (Pena and Seguin, 2001). There are two ways that transgene instability can occur—through the loss of the gene from the host or through the shutdown of expression of the gene in the host plant (gene silencing).

Transgene loss. Not all transformation leads to stable integration and inheritance of the transgene. A first step in plant transformation is to identify and discard those cells or plants in which the transgenes have been lost. Stable transformants are considered those that pass the transgene on to subsequent generations through meiosis or, in perennial plants, to be present continuously for several years (dormancy cycles). Transgenes can still be lost after several generations, however

continued

BOX 3-1 Continued

(e.g. Srivastava et al., 1996). Of the two technologies—*Agrobacterium*-mediated gene transfer and microprojectile bombardment—the latter can deliver many copies of the transgene into each recipient cell. Those multiple copies often integrate as tandem repeats, which can lead to excision (loss) of the transgenes or to gene silencing. Although *Agrobacterium*-mediated gene transfer tends to provide more stable integration, complex integration patterns can occur, including truncation of parts of the T-DNA (e.g., McCabe et al., 1999).

Gene silencing. After transformants are selected in which the transgene has stably integrated, loss of the phenotype can occur subsequently, because of the loss of expression of the transgene—by gene silencing. In the *Arabidopsis thaliana* model system, transgene inactivation has been correlated with multiple copies of the transgene, with the presence of vector backbone sequences, with DNA methylation, and with transgene position in a genome (De Buck et al., 2001; De Wilde et al., 2001; Meza et al., 2002). In *Arabidopsis thaliana* lines that contained single copies of antibody genes, De Wilde and colleagues (2001) found that silencing of transgenes can result from gene dosage effects. Homozygous lines exhibited gene silencing, and hemizygous plants showed high transgene expression. Meza and colleagues (2002), however, reported that the known mechanisms of gene silencing are not always sufficient or necessary for the induction of transgene silencing in T-DNA-transformed *Arabidopsis* lines. De Buck and colleagues (2001) showed that convergent transcription of transgenes that occur, as in an inverted repeat orientation, can trigger gene silencing in *Arabidopsis*. Highly transcribed transgenes or transgene loci that produce double-stranded RNA because of the presence of inverted repeats can result in gene silencing. Based on their observations with transgenic *Arabidopsis*, Beclin and colleagues (2002) proposed a complex pathway for RNA silencing in plants in which transgene methylation would result from production or action of dsRNA.

Post-transcriptional gene silencing induced by double-stranded RNA—termed RNAi (for RNA-interference)—occurs naturally in plants as part of a defense mechanism against virus infection. Tenllado and colleagues (2003) showed that expression of transgene constructs encoding hairpin RNA homologues can interfere with virus multiplication in a sequence-dependent manner. Double-stranded RNA was identified as the triggering structure for the induction of a specific and highly efficient RNA silencing system. The enzyme complexes facilitate the processing of dsRNA into characteristic small RNA species, known as small interfering RNAs (siRNA) that promote degradation of cognate RNAs. Tang and colleagues (2003) conducted a biochemical analysis of RNA silencing and reported that endonuclease complexes guided by small RNAs (endogenous microRNA) are a common feature of RNA silencing in animals and plants. Metzlaff (2002) showed that one component of a signal that transmits RNA silencing rapidly from silenced to nonsilenced cells by short- and long-distance signaling involves a specific, degradation-resistant RNA.

Transgene expression instability is an active area of research, with important implications for the long-term efficacy of deregulated transgenes. Some causes of transgene silencing, such as multiple copy number, can be detected easily during the early stages of development of new genetically engineered varieties. Progress is being made on new methods for detecting and reducing other causes of gene silencing. More research is needed to explain the causes of transgene instability so that researchers can develop more sophisticated techniques to minimize the problem.

and woody species because it reduces forage digestibility and is difficult to extract during pulp and paper making. The characterization of lignification genes and their exploitation for modulation of lignin profiles in transgenic plants has already been documented (Baumberger et al., 2002; Merkle and Dean, 2000; Sederoff, 1999). Boudet and colleagues (1998) reviewed the features essential to the use of transgenic approaches to lignin modification, including potential unwanted side effects and stability of transgene expression. In general, it is expected that modifications to lignin content in genetically engineered trees will either fall within the range of natural variation, which can include null mutations in key genes (MacKay et al., 1997), or will reduce fitness. The escape of those trees or their genes into wild populations would thus provide new alleles that confer no particular selective advantage and that under many circumstances would be selected against in highly competitive situations. To confirm that lignin modifications, or epigenetic changes associated with the tissue culture process, have not inadvertently improved fitness in genetically engineered tree lines, data could be gathered on growth rates and other fitness traits during the initial, limited field trials that are typically mandated for evaluation of new genetically engineered trees (McLean and Charest, 2000) before they are released. Such data on GE trees should be compared with data for nonregulated genotypes already in cultivation.

In other situations, confinement of trees could be warranted by environmental concerns. For example, it would be prudent to confine traits that disrupt the attack of plantation-grown trees by indigenous organisms that normally coexist with the tree species in natural settings and that are part of larger food chains. This would include native herbivores that rely on tree species as a primary food source and microbes that are required for mycorrhizal symbioses or nutrient recycling. It also could include resistance to exotic pests that is accomplished by transgenes that also confer resistance to nontarget native species. Genetic engineering for resistance does not in itself create risk that is much different from that attributable to resistance genes that are incorporated through multigenerational backcross breeding. However, breeding for resistance traits has not been widely attempted for forest trees in the past because of the inherent difficulties in protracted multigenerational breeding. Thus, traditional knowledge does not exist to evaluate whether gene flow from resistant genotypes in managed stands to wild stands will disrupt existing food chains or have other significant nontarget effects in subsequent generations. After field studies are done (e.g., during short-term trials, for example) for pest resistance genes bioconfinement could be considered less necessary.

Finally, when the risks associated with environmental disruption after even very limited transgene escape are too great, the release of genetically engineered trees in some locations could be deemed inappropriate, even if bioconfinement and physical confinement strategies are used in context. Hypothetical examples include the secretion of powerful allelopathic or antibiotic chemicals from the roots of those trees or the release of toxic, volatile compounds from their leaves. If the transgene were to provide a substantial competitive advantage, even severe restriction of transgene flow might be insufficient to prevent colonization after escape. Such extreme cases are trait and species specific and should be identified during risk analysis in the planning or development phases. For further discussion of the necessity of the bioconfinement of trees, see Box 3-2.

BOX 3-2
When Will Bioconfinement be Necessary for Trees?

In the past, trees were regarded as unfavorable organisms for research because of their mating systems, long life cycles, distinct juvenile-mature phases, and the fact that trees usually grow in natural settings in which genetic control of complex traits is obscured by environmental effects. However, molecular genetics, genomics and genetic engineering have opened new opportunities for research with trees. The first transgenic tree was an herbicide-resistant hybrid poplar (Fillatti et al., 1987). Genetic transformation has subsequently been applied to a variety of commercial and environmental objectives in forest trees, fruit and nut trees, and other woody perennials. Traits of interest for genetic engineering in trees include lignin (pulp) modification, increased growth and productivity, enhanced utilization of resources, pest and disease resistance, stress tolerance, herbicide resistance, optimization of mycorrhizal symbioses, phytoremediation of contaminated soils, and even production of anticancer drugs (Han et al., 1994). Genetically engineered trees are appearing with increasing frequency: Since 1989, more than 230 permits have been approved by the Animal and Plant Health Inspection Service in the United States and at least 65 have been granted in other countries for field trials on trees and other woody plants (Table 3-2). GE tree species in field tests range from pines to persimmons and poplar to papaya. Future applications of genetic engineering are likely to include restoration of species to native habitat, adaptation of trees to plantation management (domestication), enhanced fiber and biofuels production, and agroforestry.

Concomitant with the many possibilities for improvement of trees through genetic engineering are questions about the efficacy and safety of the technology in trees and other perennial plant species. The concerns raised with GE trees include the long-term stability of expression of foreign genes, the long-term effects of transgenes on nontarget species, and the long distance dispersal of transgenes through seed and pollen to wild tree stands. GE with trees presents a special challenge in that trees are often dominant species in their ecosystem and support a large web of organisms that either directly or indirectly rely on them as the ultimate source of nutrients. Many of the concerns over deployment of GE trees could be addressed with the use of appropriate bioconfinement techniques. If successful, bioconfinement could both protect the investments made in the development of genetically engineered trees and safeguard the environments in which they are grown.

continued

BOX 3-2 Continued

Initial tests of bioconfinement techniques could be conducted in fast-growing trees, such as hybrid poplar, in which testing over a full crop rotation could be as little as 6 years. In slower growing tree species, and for initial large scale tests, the testing of new bioconfinement techniques could be piggy-backed on releases of GE trees that carry transgenes of commercial importance but that are considered to pose little or no environmental or health risk. Combining bioconfinement tests with GE tree releases would provide commercial benefit upon harvest while generating important data on the effectiveness of new confinement techniques, prior to their use in situations in which successful confinement is essential.

Due to the length of time required to develop and test GE trees, decisions on whether or not to include bioconfinement methods should be made at the onset, rather than after the fact. Approaches to evaluating the need for bioconfinement are discussed in Chapters 2 and 6. The following questions could serve as a guide in the decision making process on when to incorporate bioconfinement with GE trees:

- Has it been determined previously that bioconfinement was required for similar products (the same class of genes, gene products, and vectors in similar tree species)?
- Has new empirical data been gathered to indicate that bioconfinement is still warranted?
- If this class of genes has not been evaluated in trees before, what specific, novel risk(s) might the GE tree pose to the environment, assuming certain levels of gene flow?
- Will there be closely related tree species within a distance of the site where the GE trees are grown to pose a risk of gene flow?
- If risk to the environment is considered significant, then what degree of confinement will be necessary to render the risk acceptable?
- Which currently available bioconfinement technique, if any, can provide sufficient dilution or exclusion of unwanted transgenes in other populations?
- How will the success of bioconfinement efforts be evaluated?

Risks of Most Concern with Trees

Risks Associated with Gene Flow into Natural Populations

One concern about genetically engineered trees is the consequences of gene flow from managed stands of those trees to wild populations (Slavov et al., 2002). Sexual hybridization can occur naturally between plant species that are within the same genus or occasionally, between related species in different genera. Transgene flow could be substantial if genetically engineered trees are permitted to reach sexual maturity and flower within the natural geographic range of wild relatives. The extent of hybridization between transgenic plants and their wild relatives and the scale of genetically engineered tree deployment (relative to the size of natural populations) will determine both the rate of the incorporation of transgenes into wild populations and the feasibility of confinement (Wilkinson et al., 2000). Some genetically engineered trees also could disperse seeds and establish naturalized populations.

The first step in defining appropriate confinement strategies for genetically engineered trees should be the acquisition of data on wild or naturalized tree populations (location, species mixes, distances from anticipated GEO release sites). The next step would be to assess the extent to which gene flow in pollen or seeds occurs between managed and unmanaged populations. In commercial poplar plantations, studies involving nontransgenic DNA markers have shown unexpectedly low levels of gene flow to nearby wild populations, despite the potential for extensive gene flow (Slavov et al., 2002). Slavov and colleagues (2002) hypothesize that the low gene flow observed could be attributable to the DNA marker systems used. They developed a spatial simulation model that incorporates a variety of ecological and genetic parameters to estimate the levels of future transgene flow from poplar plantations. The model permits virtual experiments to investigate how genetics, ecology, and management might influence the magnitude and variance of gene flow over 50 or 100 years. Similar models should be developed for other species and other planting scenarios, so that confinement and monitoring programs can be evaluated and designed as necessary. Case-by-case analysis will be required for useful predictions of transgene flow. Gene flow is not expected to be a problem, in and of itself, unless it leads to undesirable consequences (Box 3-1).

Effects on Nontarget Organisms

One category of objectives with the genetic engineering of trees is to prevent or reduce damage from specific pests and pathogens in tree plantations. Genes that produce pesticides with activity against groups of organisms—such as *Bt* toxin genes that protect against lepidopterans—could require confinement or restriction in expression to a much greater degree than would species-specific toxins, especially if significant gene flow is possible. Not all transgenes will affect commensal organisms, even those that have marked effects on growth. Hampp and colleagues (1996) documented that the in vitro synthesis of ectomycorrhiza between roots of transgenic aspen (*Populus tremula* × *P. tremuloides*) and *Amanita muscaria* were not affected by transformation and expression of indoleacetic acid (IAA) biosynthetic genes in roots. In contrast, Puterka and colleagues (2002) demonstrated that genetically engineering a clone of Bartlett pear, *Pyrus communis* L., (with a synthetic antimicrobial gene, D5C1) to control bacterial fireblight disease, *Erwinia amylovora* conferred unintended activity against a nontarget pest organism, the pear psylla, *Cacopsylla pyricola* Foerster. In agriculture, such cross-reactivities can be beneficial. Within natural forest ecosystems, however, they might not be.

Secondary Phenotypic Effects of Transgenesis

As in nontransgenic methods of genetic modification, the process of inserting transgenes into plant cells and regenerating whole plants from those cells can result in different types of unintended phenotypic effects. Those "secondary" phenotypes can be caused by the transgene's location in the genome, by effects of the transgene on other traits (pleiotropy), by interactions among the transgene and native genes (epistasis), and by somaclonal mutations that occur during tissue culture. Carefully designed control lines can identify specific causes of unintended phenotypes, but such detailed analyses often are lacking. Several examples of unintended phenotypes in trees are described below. It is important to note that their range is expected to be much smaller in deregulated plants that have been extensively evaluated in field trials than in experimental lines used in research. It is also important to note that unintended effects are not necessarily undesirable.

Ralph and colleagues (2001) reported on the production of unanticipated benzodioxane structures in lignins of transgenic poplar plants deficient in COMT, an O-methyltransferase required to produce lignin syringyl units. This demonstrates the ability of plants to accommodate mutations in gene expression that might not be predicted based on current knowledge of biochemical pathways. Holefors and colleagues (2000) constructed transgenic apple rootstock clones that carried 1–8 copies of the *Arabidopsis* phyB gene. Multiple effects of phyB overexpression were observed, including reduction in stem length in 9 of 13 clones and reduction in shoot, root, and plant dry weights in all transformed clones compared with untransformed control plants. Atkinson and colleagues (2002) produced transgenic apple (*Malus domestica* Borkh. cv Royal Gala) trees that contained additional copies of a fruit-specific apple polygalacturonase gene (PG) under a constitutive promoter. In previous studies in transgenic tobacco (*Nicotiana tabacum*), PG overexpression had no effect on the plant, but in the transgenic apple it led to a range of phenotypes, including silvery colored leaves and premature leaf shedding. Mature leaves had malformed and malfunctioning stomata that perturbed water relations and contributed to a brittle leaf phenotype. O'Connell and colleagues (2002) produced transgenic tobacco plants that severely suppressed the activity of cinnamoyl-CoA reductase (CCR) as a model for altering lignin content in trees. Although transgenic lines had the desired changes in lignin structure, some showed a range of aberrant phenotypes, including reduced growth.

Other pleiotropic effects can help a plant. Hu and colleagues (1999) observed that transgenic aspen (*Populus tremuloides* Michx.) with down-regulated expression of the gene that encodes 4-coumarate:coenzyme A ligase (4CL) exhibited a 45% reduction in lignin and a 15% increase in cellulose. The total lignin–cellulose mass in the genetically engineered trees was essentially unchanged. Furthermore, leaf, root, and stem growth were substantially enhanced, and structural integrity was maintained both in the cells and in whole plants in the transgenic lines. Those results indicate that metabolic flexibility can sustain the long-term structural integrity required of woody perennials in transgenics. El Euch and colleagues (1998) transformed walnut (*Juglans nigra* × *Juglans regia*) with an antisense chalcone synthase (chs) gene that not only reduced flavonoid content in the stems of the plants but also enhanced adventitious root formation.

Some studies have been designed to generate pleiotropic effects intentionally, such as the introduction of oncogenes (rolC) from *Agrobacterium*. Grunwald and colleagues (1999, 2001) compared the morphology, wood structure, and cell wall composition in rolC transgenic hybrid aspen (*P. tremula* × *P. tremuloides*) with nontransformed control trees. The transgenic trees had stunted growth, altered physiological parameters, and light green leaves that were smaller than normal. Numerous alterations also were observed in the formation and differentiation of xylem cells. In contrast, when Tzfira and colleagues (1999) expressed rolC transgene in aspen (*Populus tremula*) they observed both accelerated growth and improved stem production index in the transgenic plants. Eriksson and colleagues (2000) produced transgenic hybrid aspen (*Populus tremula* × *P. tremuloides*) overexpressing a key regulatory gene in the biosynthesis of gibberellin (GA). The transgenic trees had improved growth rate and biomass, as expected, but they also had more numerous and longer xylem fibers than did wild-type plants.

Not all engineered mutations in trees produce such secondary effects, of course. Bhatnagar and colleagues (2001) altered polyamine metabolism in cells of transgenic poplar (*Populus nigra* × *P. maximowiczii*) by expressing a mouse Orn decarboxylase (odc) cDNA. The transgenic cells showed the expected effects on polyamines, but the overall arginine pathway was not affected and assimilation of nitrogen into glutamine kept pace with the increased demand for putrescine. Transgenic citrus seedlings that constitutively expressed the LEAFY (LFY) or APETALA1 (AP1) genes from *Arabidopsis* showed dramatically precocious flowering and fruit production, as desired, without exhibiting any other developmental abnormalities (Pena et al., 2001).

Instability of Transgene Expression

Pena and Seguin (2001) pointed out that stable integration and expression of foreign genes is particularly important for long-lived species such as trees. Instability that is not detected prior to field release is clearly undesirable.

A better understanding of the causes of transgene expression instability can help researchers develop more sophisticated techniques to minimize the problem. To investigate transgene silencing in a tree system, Kumar and Fladung (2001) analyzed aspen (*Populus tremula* L.) and aspen hybrid (*P. tremula* L. × *P. tremuloides Michx.*) lines transformed with the rolC phenotypic marker system and grown in vitro, in greenhouses and in the field. Their molecular analyses showed that, in the hybrid aspen genetically engineered lines, the inactivations were always a consequence of transgene repeats (multiple incomplete or complete copies). In wild nonhybrid aspen, however, instability in some of the transgenic lines was the result of a position effect. This indicates that the tree host genome has some control over expression of transgenes. Other studies have shown that expression of transgenes in poplar can be stable under field conditions in many cases (e.g. Meilan et al., 2002).

Dominguez and colleagues (2002) studied transgene silencing in Mexican lime (*Citrus aurantifolia* [Christm.] Swing.) transformed with the *Citrus tristeza* virus coat protein gene. More than 30% of the transgenic limes that had regenerated under nonselective conditions exhibited silencing of the transgenes. They observed that inverted repeats as well as direct repeats and even single integrations triggered gene silencing. In contrast, Cervera and colleagues (2000) studied 70 transgenic citrus plants in a screenhouse over 4–5 years. They observed only 4 phenotypic off-type plants, all of which were the result of tetraploidy in the tissues used for transformation. Gene-silencing or pleiotropic effects were not to blame.

Options and Constraints

Sterility

The creation of sterile trees has attracted wide support and interest as a method of bioconfinement. Strauss and colleagues (1995) discussed the regulatory and ecological rationales for engineering sterility in trees, the strategies for creating sterility-inducing transgenes, and the problems peculiar to engineering sterility in forest trees. Two primary options for genetically engineering sterility in trees are being pursued: ablating floral tissues through floral-specific promoter–cytotoxin fusions and disrupting expression of essential floral genes by gene suppression (Strauss et al., 1995). Both options should be thoroughly tested, as each has advantages and disadvantages. There are other approaches for generating sterility that do not require genetic engineering. Cytoplasmic male sterility can be created through nuclear-cytoplasmic (mitochondrial) incompatibilities resulting from the crossing (sexual hybridization) of specific genotypes or related species (e.g. Shi and Hebard, 1997). Sterility has been observed to occur naturally in some tree species (e.g. Linares, 1985; Soylu, 1992) and has been selected in ornamental trees for which fruits are not desirable or when seedless (parthenocarpic) fruits are preferred (Rapoport and Rallo, 1990; Talon et al., 1992).

As in so many aspects of designing safe genetically engineered trees (Pena and Seguin, 2001), long life span can create problems for the use of sterility as the sole bioconfinement tool. The systems developed for sterility in trees must be highly stable if they are to be trusted in cases where rotation lasts for 40 years or more. Stable suppression of fertility could require targeting of multiple floral genes or the combined use of several genetic mechanisms for inducing sterility and other bioconfinement methods. Engineering of complete sterility or male sterile lines could help to achieve gene confinement, and it could stimulate faster wood production, reduce the production of allergenic pollen, and (in the case of male sterility) facilitate hybrid breeding in trees.

Triploidy

Results of controlled crosses have shown that triploid clones used in genetic engineering experiments with poplar have a high level of innate sterility and are less competitive in mixed stands than are their wild relatives (Strauss and Meilan, 1997). Such chromosomal abnormalities could provide natural systems for bioconfinement. The stability of sterility in such lines must be evaluated, and an expected frequency of somatic reversions to fertility should be determined, or at least estimated, over the expected rotation for each tree crop before incorporation into confinement strategies.

Gene Silencing

RNA silencing can be induced in plants by several mechanisms including induction of effects associated with transcriptional, posttranscriptional, genomic DNA methylation, and gene dosage events. In trees and other perennial plants, gene silencing occurs naturally (Fraga et al., 2002) and transgenically (Dominguez et al., 2002; Kumar and Fladung, 2001). Initially, posttranscriptional silencing was accomplished with antisense or cosuppression. These constructs usually result in only a modest proportion of silenced individuals, however, which is not useful for long-term field trials of trees. The steps involved in gene silencing in plants are becoming better understood and should lead to the development of improved transgene bioconfinement tools for trees. If successful and durable under field conditions, gene silencing could find broad application in the improvement of transgenic trees and rootstocks.

Fitness Handicaps

Bioconfinement by auxotrophy involves the use of natural or transgenic mutant genotypes that lack a necessary biosynthetic component. Tree mutants would be rendered much less competitive than their wild relatives in unmanaged forest environments, or they would simply not survive, assuming that the missing biosynthetic component was not available to escaped plants and that the mutations could not be complemented (recovered) in hybrids with wild-type trees. The novel recoverable block of function (RBF) technique of Kuvshinov and colleagues (2001) is an engineered form of auxotrophy that could reduce gene flow from transgenic trees to wild relatives. The RBF system is superior to single-gene-mutation auxotrophs because hybrids between the transgenic plants carrying the RBF and the wild relatives would die or be unable to reproduce because of the blocking construct. In practice, most forms of auxotrophy would be difficult to apply given the recessive nature of most auxotrophic mutations, the slow nature of the process to create homozygous individuals, and the high degree of inbreeding depression shown by trees. The large-scale chemical applications that are required for some auxotrophs could lead to ecological harm. However, use of the RBF technique for engineering conditional auxotrophy could be an effective way to confine transgenes and should be investigated more thoroughly for trees.

Tissue-specific Expression

It is often desirable to restrict the expression of transgenes in genetically engineered plants to the tissues that require the encoded activity. Promoters from genes in the lignin pathway have been demonstrated to impart tissue- and development-specific expression of marker genes in transgenic plants. The development of such "regulated" promoters opens the possibility for restricting expression of economically important transgenes in trees to those situations in which their gene products are required. In the case of resistance genes, such specificity could lessen the selection pressure on the disease and pest populations and help avoid breakdown of resistance over the many years that a transgenic tree could grow. Regulated expression of transgenes for biotic resistance also will help lessen effects on nontarget organisms.

The transformation of poplar is often used to examine the roles of upstream and downstream regulatory elements from other tree genes. Two loblolly pine genes that were developed for xylem cDNA libraries exhibited tissue specificity effects in leaves of transgenic poplar that were promoter specific (No et al., 2000). Gray-Mitsumune and colleagues (1999) demonstrated the specificity of expression of the poplar PAL promoter to vasculature in both transgenic poplar and spruce. Genome-sequencing projects, such as the poplar genome sequencing project (http://www.ornl.gov/ipgc/), coupled with global analysis of gene expression studies, will unveil thousands of regulated promoters that will be tested for use in transgenic constructs in studies of functional genomics.

Plastid Engineering

As an alternative to nuclear transformation, transgene expression from the chloroplast genome offers several advantages, including high-level foreign protein expression and lack of pollen transmission for improved transgene confinement in angiosperms (reviewed by Bock, 2001). Daniell and colleagues (1998) first reported genetic engineering by stable integration of a foreign gene into the tobacco chloroplast genome. Improved chloroplast-based expression systems now include vectors, expression cassettes, and site-specific recombinases for the selective elimination of marker genes (Maliga, 2002). Because chloroplasts are transmitted through pollen in gymnosperms, the chloroplast transformation approach to bioconfinement is limited to hardwood tree species. An analogous system for bioconfinement of foreign genes in conifers would be mitochondrial transformation, although stable, efficient mitochondrial transformation has not yet been reported. Given the extensive dispersal rates of genes through tree pollen, more effort should be placed on chloroplast transformation for bioconfinement in hardwood trees than has been reported to date, especially if sterility is not an alternative. Seed transmission of transgenes is also of more concern for tree species than it is for annual crops. Thus, chloroplast transformation would need to be combined with other methods to achieve strict bioconfinement in trees.

Outlook for Bioconfinement of Transgenes in Trees

Safe and effective bioconfinement methods should lead to greater acceptance of transgenic trees by the general public and to increased opportunities for the creation and deployment of genetically engineered trees by researchers in the public and private sectors. Given the number of options available and the frequency with which new approaches are being reported, there is every reason to believe that effective bioconfinement methods will soon be available for trees. Whether the full development, testing, and deployment of bioconfinement methods, as well as the application of genetically engineering itself, will be realized for trees is still open to question, however. This will only occur if there is acknowledgment by industry that bioconfinement of transgenes is necessary and beneficial; if the public accepts that there can be an acceptable risk; and if more public funding becomes available for the discovery, development, and appropriate testing of bioconfinement methods in trees.

The likelihood of public acceptance of nominal risk associated with the growth of trees under bio- and physical confinement could be improved by the use of transgenes derived from the same or similar tree species and by the development and adoption of methods for monitoring wild populations for entry of transgenes from genetically engineered tree plantations. Those steps are already being pursued to a limited extent. For example, Marcus and colleagues (1997) isolated the gene for an antimicrobial peptide (MiAMP1) from the nut kernels of *Macadamia integrifolia* that inhibits the growth of fungal, oomycete, and gram-positive bacterial phytopathogens in vitro, but that is nontoxic to plant and mammalian cells. Such genes could prove as useful in genetic manipulations to increase disease resistance in transgenic trees as would be genes derived from bacteria. Connors and colleagues (2002) are studying whether modification of a cystatin gene isolated from American chestnut (*Castanea dentata*) could confer chestnut blight resistance.

Diagnostic tests for transgenes can be applied readily to genetically engineered trees for determining the location of transgenes, as described above. However, because natural forests often cover such large areas it would be difficult to monitor them effectively on a random basis for transgene movement. Wilkinson and colleagues (2000) used remote sensing to identify sites of sympatry between *Brassica napus* and its progenitor species across 15,000 km^2 of southeastern England before the release of transgenic *B. napus* plants. This work allowed the researchers to focus their activities in areas where transgene escape was most likely. The same approach could be taken to identify the best sites for release of genetically engineered trees and to determine where monitoring should occur. Efforts are under way to develop remote sensing that detects expression of transgenes based on the unique profile of volatile compounds that can serve as signatures for genetically engineered plants (www.aginfo.psu.edu/News/march03/sentinel.html). Ghorbel and colleagues (1999) have shown that green fluorescent protein (GFP) can be used as a visible marker for selection of transgenic woody plants, as an alternative to antibiotic and herbicide selection. GFP might thus serve as a marker for monitoring trees for genetic escape through the use of remote or handheld ultraviolet (UV) light sources or fluorescence detectors.

TRANSGENIC GRASSES

Many types of grasses have been considered for genetic engineering, including forage grasses for rangeland and native grasses for biomass production or bioremediation (NRC, 2002a; Wipff and Fricker, 2001). The most advanced and most profitable species, however, are turfgrasses, and are used widely in landscaping and on golf courses, for example. There are more than 14,000 golf courses, 40,000 athletic fields, and 40 million parks and home lawns in the United States (Edminster, 2000). And the U.S. turfgrass seed market is second only to the hybrid corn seed market, with annual sales of $580 million to $1.2 billion (Wipff and Fricker, 2001). There is considerable interest and investment in turfgrass science, much of it supported by the United States Golf Association (Kenna, 2000).

Genetically Engineered Turfgrasses

Since 1993, the Animal and Plant Health Inspection Service has issued more than 200 permits for small field tests of transgenic turfgrass species in the United States (Table 3-3; Wipff and Fricker, 2001), although none has yet been deregulated by the U.S. Department of Agriculture. Monsanto's glyphosate-resistant creeping bentgrass (*Agrostis palustris*; http://gophisb.biochem.vt.edu) was the first transgenic grass to be considered for approval. Monsanto also is developing a new lawn grass that requires less mowing than does its nontransgenic counterpart.

Although the development of transgenic grasses has fallen behind the research being done on major crop plants, a great deal of basic research has addressed transgenic methods for improving turfgrass cultivars. Biolistic bombardment, protoplast DNA uptake (either through electroporation or mediated by polyethylene glycol), and recently *Agrobacterium*-mediated transfer of genes have been used for genetic transformation of turfgrasses. Techniques for in vitro regeneration also have been developed (e.g., Chai and Sticklen, 1998; Lee, 1996). Biolistic gene bombardment of creeping bentgrass (*Agrostis palustris* Huds.) with the reporter *gus* was developed, and a GUS enzyme histochemical assay was used to identify nonchimerically transformed plants (Zhong et al., 1993). Biolistic bombardment of turfgrass callus or suspension cells, and electroporation-mediated or polyethylene-glycol-mediated protoplasts have been used to transfer hygromycin, phosphinothricin, biolophos, NPTII, or G148 resistance selectable marker genes in stolonate bentgrass (*Agrostis stolonifera* var. palustris) (Sugiura et al., 1997, 1998), creeping bentgrass (*Agrostis palustris* Huds.) (Hartman et al., 1994; Lee et al., 1996; Liu, 1996, Zhong et al., 1993, 1998), red top (*Agrostis alba* L.) (Asano and Ugaki, 1994), orchardgrass (*Dactylis glomerata* L.) (Horn et al., 1998), tall fescue (*Festuca arundinaceae* Schreb)

TABLE 3-3 Genetically Engineered Turfgrasses, Permits Approved by APHIS for Field Tests in the United States, 1993–2003

Organism	Phenotypes (Number of Permits)	Gene
Bermudagrass	AP - Drought and salt tolerance increased (9)	LEA or Lsd or BADH
	HT - Phosphinothricin tolerant (1)	PAT
Creeping bentgrass	AP - Aluminum tolerant (2)	CISY
	AP - Drought tolerant (3)	LEA & OSTL & PAT; or THIL
	AP - Drought and salt tolerance increased (9)	LEA or Lsd or BADH
	AP - Growth rate altered (3)	CBI
	AP - Growth rate altered & HT - Glyphosate tolerant (4)	CBI and/or EPSPS
	AP - Heat tolerant (2)	THIL
	AP - Salt tolerance increased (7)	BADH or LEA
	FR - Brown spot and dollar spot resistant (1)	PAT; PI-II
	FR - Dollar spot resistant (21)	TMK or AVP or GOX or BAC
	FR - *Fusarium* resistant (1)	Bcl-xl
	FR - *Rhizoctonia solani* resistant (4)	CBI
	FR - *Rhizoctonia solani* & *Sclerotinia* resistant (1)	AVP; BAC;FAD; GOX; IMT; TMK
	FR - *Sclerotinia* resistant & HT - Glyphosate tolerant (4)	CBI
	HT - Glyphosate tolerant (50)	CBI or EPSPS
	HT - Phosphinothricin tolerant (25)	PAT or CBI
	IR - Sod web worm resistant (2)	CBI
	MG; SM - Hygromycin tolerant (4)	GUS; HYR
	MG; SM - Spectromycin resistant (1)	CBI
Festuca arundinacea	FR - *Rhizoctonia* resistant (2)	AGLC; Npr1; HYR; CHT
	HT - Phosphinothricin tolerant (2)	PAT
	MG; SM - Hygromycin tolerant (3)	HYR; GUS
	PQ - Lignin decreased (3)	CAD and/or COMT, COMT antisense
Kentucky bluegrass	AP - Drought tolerant (1)	BADH
	AP - Drought and salt tolerance increased (8)	Lsd or BADH or LEA
	AP - Growth rate altered (1)	CBI
	AP - Growth rate altered & HT - Glyphosate tolerant (6)	CBI
	FR - *Rhizoctonia solani* resistant (2)	CBI
	HT - Glyphosate tolerant (6)	CBI

TABLE 3-3 Continued

Organism	Phenotypes (Number of Permits)	Gene
Paspalum notatum	PQ - Lignin decreased (1)	COMT; PAT
Perennial ryegrass	AP - Drought and salt tolerance increased (3)	BADH; HYR
Poa pratensis × *Poa arachnifera*	AP - Growth rate altered & HT - Glyphosate tolerant (1)	CBI
	HT - Glyphosate tolerant (2)	CBI
Russian wildrye	MG; SM - Hygromycin tolerant (3)	GUSi; HYR
St. Augustine grass	AP - Growth rate altered & HT - Glyphosate tolerant (6)	CBI
	HT - Glyphosate tolerant (6)	CBI
	HT - Phosphinothricin tolerant (1)	CBI
Velvet bentgrass	HT - Phosphinothricin tolerant (1)	PAT

NOTE: APHIS; Animal and Plant Health Inspection Service. Field test data downloaded from Information Systems for Biotechnology, http://www.nbiap.vt.edu/cfdocs/fieldtests2.cfm, May 23, 2003 does not include submissions denied or withdrawn.

Phenotype Key: AP, agronomic properties; FR, fungal resistance; HT, herbicide tolerant; IR, insect resistant; MG, marker gene; PQ, product quality; SM, selectable marker.

Gene Key: AVP, antiviral protein from pokeweed ; BAC, bacteropsin gene from *Halobacterium halobium*; BADH, Betaine aldehyde dehydrogenase from garden orach (*Atriplex hortensis*); Bcl-xl, B-cell lymphoma related gene from chicken; CAD, cinnamyl alcohol dehydrogenase from tall fescue; CBI, confidential business information; CHT, chitinase from rice; CISY, COMT, caffeate O-methyltransferase gene from tall fescue; EPSPS, 5-enolpyruvylshikimate-3-phosphate Synthase; FAD, delta-9 desaturase from *Saccharomyces cerevisiae* ; GOX, glucose oxidase from *Aspergillus niger*; GUS: *E. coli* β-glucuronidase gene; HYR, hygromycin phosphotransferase gene; IMT, inositol methyl transferase; LEA, late embryogenesis abundant protein gene from barley; Lsd (Sac), levansucrase gene from *Bt*; Npr1, nonexpressor of pathogenesis-related gene from *Arabidopsis thaliana*; NPTII, neomycin phosphotransferase gene from *E. coli*; OSTL, thaumatin-related protein from rice; PAT, phosphinothricin acetyl transferase gene from *Strep. hygroscopicus*; AGLC, β-1,3-glucanase antisense from alfalfa; PI-II, proteinase inhibitor II from potato; SAR, systemic acquired resistance gene from *Arabidopsis thaliana*; THIL, thiamine biosynthetic enzyme from corn; TMK, receptor kinase gene from *Arabidopsis thaliana*.

(Dalton et al., 1995; Ha et al., 1992; Spangenberg et al., 1994; Wang et al., 1992), red fescue (*Festuca rubra* L.) (Spangenberg et al., 1994), perennial ryegrass (*Lolium perenne* L.) (Alpeter et al., 2000; Spangenberg et al., 1995), Italian ryegrass (*Lolium multiflorum*) (Potrykus et al., 1985), Japanese lawngrass (*Zoysia japonica* Steud.) (Inokuma et al., 1997, 1998), and European turf type red fescue (*Festuca rubra* L.) (Alpeter et al., 2000). More recently, the *Agrobacterium*-mediated genetic transformation system was used to transfer the *gus* reporter gene and the hygromycin phospho-transferase (*HTP*) genes in Korean lawngrass (Chai et al., 2000) and the GFP and *HTP* genes in creeping bentgrass (*Agrostis palustris* L.) (Chai et al., 2000).

Potential for Gene Flow

Grasses that are cultivated for turf, forage, and ornamental uses pose several challenges for bioconfinement because of their capacity for out-crossing, hybridization, and vegetative propagation. Also, many cultivated grasses are closely related to noxious weeds. Turfgrasses are open-pollinated plants that often cross-pollinate with weedy species. For example, the bentgrass group (*Agrostis*) has more than 100 species, many of them weeds that can hybridize with one another (Wipff and Fricker, 2001). Bermuda-grass (*Cynodon dactylon*. L. var. Prsoon) is an important perennial forage and turfgrass that is considered to be a weed in many regions of the United States. Should bermudagrass be genetically engineered to improve hardi-ness, the escape of transgenes has the potential to cause ecological and economic damage (Ellstrand and Hoffman, 1990). Giddings and colleagues (1997a) noted that, in the United Kingdom, forage grasses including *Lolium perenne* cultivars are sexually compatible with wild and feral species of the same genus and with fescue species (*Festuca* spp.; Figure 3-3). *Agrostis* species can hybridize with *Polypogon* species, and it is believed that *Agrostis parlatore* and *A. moldavica* are derived from past hybridization between *Agrostis casstellana* and *Polypogon viridis* (Wipff and Fricker, 2001). Some commercially important grass species can hybridize with nearby congeners and then switch to asexual seed production (apomixis), allowing crop genes to spread widely even when F_1 hybrids are sexually sterile (Wipff and Fricker, 2001). Because turfgrasses are perennial, the longevity of unintended perennial hybrids between transgenic and wild plants will increase the opportunities for further backcrossing with other wild or domestic grasses.

The rate of introgression of some turfgrasses is actually higher for interspecific and intergeneric hybrids than it is among intraspecies crosses (Wipff and Fricker, 2001). For example, creeping bentgrass is self-incompatible (self-sterile) but highly cross-compatible with other species (Bjorkman, 1960). An early report detected, through paternity analysis, more than 1%

FIGURE 3-3 A wild hybrid, *F. arundinacea* and *L. multiflorum* Lam.

cross-pollination of nontransgenic creeping bentgrass plants at a distance of 8,000 m (Ellstrand and Hoffman, 1990). Turfgrasses have small pollen that can blow great distances. Normally, the two factors of distance and wind direction are considered to predict the distance that pollens can travel (Giddings, 2000; Giddings et al., 1997b). However, other factors, such as speed and wind turbulence—especially if "whirl winds" are present—are important in the unintended deposition of pollen in other fields. Other factors include relative humidity and temperature (Wipff and Fricker, 2001). Because there are no models to predict those factors, an old method of exponential power function (Bateman, 1947) can be used to predict turfgrass pollen disposition (Wipff and Fricker, 2001).

Wipff and Fricker (2001) measured gene flow from herbicide-resistant transgenic creeping bentgrass into wild relatives. The primary objectives of the study were to investigate intra- and interspecific gene flow of transgenic creeping bentgrass in the Willamette Valley of Oregon, where nearly all U.S. bentgrass seed is produced. Pollen movement was determined by placing transects of nontransgenic creeping bentgrass around a nursery of 286 plants genetically engineered for tolerance to the herbicide glufosinate. In 1998, transgenic turfgrass pollen grains were observed to travel 1,066.8 m along southwest transects and 1,309.4 m along northeast transects from the nursery. In 1999, transgenic pollen traveled 331.5 m to the southwest, 575.1 m to northeast, 262.4 m to the northwest, and 331.5 m to the southeast from the nursery. The experiments resulted in the introgression of the bar gene from creeping bentgrass into *A. canina, A. capillaris, A. castellana, A. gigantea,* and *A. pallens* species.

Turfgrasses can vegetatively multiply easily and effectively by rhizomes and stolons. Those underground parts often are translocated by machinery. Birds and mammals also facilitate the dispersal of turfgrass because they feed and forage in and around turfgrass stands for seeds and insects. Grass seeds are ingested and excreted or carried on fur or feathers for deposition elsewhere.

For all of the reasons discussed above, transgenic turfgrasses, perhaps especially creeping bentgrass, can be considered potentially difficult to confine (Box 3-3). It also must be recognized that bentgrass is a commercially important turfgrass because of its extensive use in golf courses: More than 65% of the transgenic field test permits issued have been for bentgrass (Table 3-3).

Bioconfinement Methods for Transgenic Turfgrasses

Each bioconfinement technique discussed above could be used in future transgenic turfgrass products. The possibilities include chloroplast trans-

---

BOX 3-3
Turfgrass Might be Difficult to Confine

Transgenic turfgrasses carry a particularly high risk of escape for two reasons: Turfgrasses are perennial, so they have many seasons in which to spread through pollen and seeds, and they form unintended hybrids (which themselves would be long-lived) easily. Turfgrasses are open-pollinated plants with a very high cross-ability, primarily with species that are aggressive weeds. Most turfgrasses have many species that outcross heavily among themselves (Giddings et al., 1997a) and even among different turfgrass genera. For example, in nature, *Agrostis spp.* (bentgrass) cross-breeds with members of the *Polypogon* genus; and it is believed that *Agrostis parlatorei* Breistr and *A. moldavica* Dobrescu and *A. moldavica* Beldie are derived from multiple cross-hybridization between *A. casstellana* and *P. veridis* (Wipff and Fricker, 2001). Also, there are several examples of anthropogenic hybrids between ryegrass (*Lolium spp.*) and Fescue (*Festuca spp.*) genera. Figure 3-3 shows a wild hybrid between tall fescue (*F. arundinacea*) and annual ryegrass (*L. multiflorum* Lam) developed by Tim Phillip at the University of Kentucky. More intensive bioconfinement methods, such as the use of plastid transgenesis and male sterility are needed in genetically engineered turfgrass production.

genesis, tissue- and organ-specific gene expression, male sterility, apomixis, terminator gene technology, gene silencing, suicide genes, ablation, excision, and inducible promoters. However, few bioconfinement techniques have been reported for turfgrasses, in part because little funding has been available for basic research. A significant increase in support will be needed to promote development of an adequate arsenal of bioconfinement techniques for the safe use of transgenic turfgrasses.

It should be noted that some transgenes could have beneficial effects, should they transfer to other grasses through pollen flow or by other means. Many people suffer from ryegrass pollen allergies, and ryegrass was recently genetically engineered with an antisense-mediated silencing of the gene (lot p5) that encodes the rye pollen allergen. The lot p5 gene antisense construct was expressed in ryegrass under regulation of a pollen-specific promoter. The pollen from those transgenic plants showed low IgE antibody-binding capacity of pollen extract as compared with control pollen, meaning that the pollen of the genetically modified ryegrass could contain minimal amounts of allergen or none at all (Bahalla et al., 1999). This could be of great benefit to allergy sufferers.

TRANSGENIC ALGAE

Microscopic and macroscopic algae are a diverse group of organisms that are taxonomically distinct from plants. Microalgae are discussed along with bacteria and other microbes in Chapter 5. Commercial production of macroalgae is an important sector of aquaculture, especially in Asia. Seaweeds, such as *Laminaria, Porphyra, Undaria,* and *Graciliaria,* are grown for food and food additives, including polysaccharides such as carageenan (Renn, 1997). Commercial transgenic macroalgae have not been developed, in part because of technical obstacles, but there is increasing interest in using them to enhance fuel, polysaccharide, fish feed, and pharmaceutical production and in environmental bioremediation (Minocha 2003; Stevens and Purton, 1997). As with grasses and trees, some commercially grown algae have tremendous potential to disperse and persist in natural habitats.

Some algae are considered invasive because they out-compete native species and dominate marine ecosystems when introduced to new areas (Occhipinti-Ambrogi and Savini, 2003). Because algae often are cultured outside their native ranges, some nontransgenic species have been managed using bioconfinement methods. For example, a "biological design" method has been used in Maine to confine nonengineered nori (*Porphyra* spp.). An introduced species of nori (*P. umbilicalis*) is cultivated commercially on rafts that float in coastal waters where a closely related native species of nori also occurs. Concerns were raised that the introduced species would become invasive and harm native populations by hybridization or competition. However, extensive field studies documented that, under ambient conditions, the introduced species was not invasive and did not reproduce, most likely because of its poor survival in winter (Levine et al., 2001). Thus, this nonnative nori appears to be biologically confined, as long as its reproductive capacity continues to be inhibited by local conditions.

Other bioconfinement methods would be needed for genetically engineered algae that can survive and spread in natural habitats near aquaculture facilities. There is no feasible method of inducing sterility in algae, and the lack of basic understanding of the biology of reproduction in most algae is a major obstacle to developing a feasible method in the near future. Macroalgae are plastic in growth form. They often have complex life histories that involve multiple reproductive pathways, including parthenogenesis and vegetatively dispersed propagules. Researchers do not fully understand sex determination, reproduction, or other aspects of the life history of many species; in some cases, they have not even identified which life stage is reproductive. Therefore, any efforts to study and then biologically confine transgenic algae will have to proceed on a case-by-case basis.

EFFECTIVENESS AT DIFFERENT SPATIAL AND TEMPORAL SCALES

Most of the bioconfinement methods discussed here are equivalent to natural mechanisms of reproductive isolation that act to maintain species barriers. In plants, the leakiness of those species boundaries is well known (Arnold, 1997; Grant, 1981; Levin, 1978). Within species, distinctive breeding systems such as dioecy (male or female plants) and self-incompatibility also are known to be leaky (e.g., Lloyd, 2000; Poppendieck and Petersen, 1999). Moreover, experience suggests that sterility is rarely absolute. Thus, in most circumstances, single-method efforts at bioconfinement are likely to be less than 100% effective in preventing the escape of transgenes, especially if large numbers of plants are involved. The same could be true of multiple-method bioconfinement efforts if there is a chance that individual methods could fail. Unless a bioconfinement method is 100% effective in preventing the movement of seed, pollen, spores, and vegetative propagules, its efficacy generally would vary considerably over different spatial and temporal scales.

Spatial Scale

Bioconfinement generally will work best for small numbers of plants that are physically isolated (on the order of kilometers at least) from other populations of the same species or from compatible relatives. Relatively small plant populations tend to be gene flow sinks rather than gene flow sources. All other things being equal, when population sizes vary, gene flow tends to be asymmetric: There is more flow from large populations into small ones than the other way around (Handel, 1983; Levin and Kerster, 1975). Thus, if a bioconfined crop were planted in the midst of other varieties of the same species (e.g., maize grown in Iowa), the percentage of efficacy of less-than-perfect bioconfinement would be expected to drop radically as the number of bioconfined plants increased from dozens to thousands. First, the chance of genetic changes that "disarm" confinement traits, such as mutations that silence transgenic sterility systems, increases with population size. Second, larger populations are more likely to disperse pollen, seeds, or vegetative propagules than are small populations (e.g., Handel, 1983; Levin and Kerster, 1975), and this could compromise back-up strategies such as physical isolation of the bioconfined crop. Although most of the data that associate population size and gene flow come from the literature on pollen flow, there is every reason to assume that similar relationships would occur for the dispersal of seed and vegetative propagules.

Small populations could be common for a few types of transgenic crops—such as pharmaceutical-producing plants—that are grown commercially. The high economic value of those crops and the requirement to segregate them from related crops or wild species will mandate their cultivation in small or isolated populations. However, most plants grown for other uses are likely to be cultivated on a much larger scale. If, for example, bioconfinement is desired for corn or tobacco varieties that produce industrial chemicals, some of those crops could be grown on thousands of acres with millions of plants at each site and millions of other, nontransgenic, plants growing nearby.

Another aspect of spatial scale is the number of populations that will be cultivated and the number of regions in which the crop can be grown. Local varieties of corn and soybean are grown over vast areas in the United States; fruit orchards and vineyards tend to be smaller and more regional. Major commodity crops that constitute the basis of industrialized agriculture could pose the greatest challenges for bioconfinement because they are grown on an enormous scale. Likewise, forage crops planted on rangeland occupy vast geographic areas, especially in the western states. Even highly managed tree plantations and golf courses represent large populations, each of which consists of thousands or millions of individual plants. When bioconfined plants are grown in many regions, there is a greater chance that they will be planted in the proximity of sexually compatible cultivars or wild relatives. This magnifies the chances of unwanted effects should bioconfinement break down.

Temporal Scale

In the same vein, the efficacy of bioconfinement should decrease as temporal scale increases. The longer a population is in place, the greater the chance that bioconfinement will erode, and the more opportunities the population will have to disperse pollen, seed, and vegetative propagules. Perennials are long-lived by definition, but even annual plants can occur in long-lasting populations. Indeed, if some small amount of viable seed is released undetected into the soil, that seed bank can grow considerably over a series of years. Environmental conditions also vary from one year to the next, and the efficacy of bioconfinement varies under different environmental conditions; opportunities for failure increase over time.

Perennials such as turfgrasses and trees can behave very differently from annual crops. Where annuals grow, flower, set seed, and die within a single year, perennials are heterogeneous. Depending on the species, they might or might not flower within a year of germinating. Some species do not flower for many years. Some perennial species live a few years; others (including some grasses and trees) can live for hundreds or even thousands of years. Many perennials (especially grasses) reproduce vegetatively, many do not. Each combination of species-specific temporal patterns will have a different influence on bioconfinement strategies. A perennial in which flowering is delayed for many years and in which vegetative reproduction does not occur will be relatively easy to confine, especially if plants are harvested thoroughly before they flower. At the other extreme, a perennial that creates vegetative propagules regularly, flowers at an early age, and continues to flower every year could be structured to produce so many progeny by seed, pollen, and propagule that finding an effective bioconfinement strategy could be a significant challenge.

MONITORING AND MANAGING CONFINEMENT FAILURE

The degree to which failed confinement can be monitored and managed depends on whether the GEOs are easily detected, the scale at which they are released into the environment, and their subsequent population dynamics and the degree to which they can hybridize with related species. Early detection of failed methods is important, especially if the confined transgenes are likely to spread, but this might be possible only for small-scale plantings of some crops. If a failed bioconfinement method can be recognized by distinctive phenotypic traits, such as the presence of flowers in otherwise sterile plant varieties, it might be possible to cull abnormal plants in small fields. That practice is used in certified seed production programs, where inspectors go through the fields to remove or cut off any "off-type" plants that do not conform to desired phenotypic standards.

However, failures of many bioconfinement methods will be much more difficult to detect. Elaborate experiments would be needed to identify the proper functioning of a repressible seed-lethal transgene. And most bio-confined plants will be grown on such large areas of land that repeated, comprehensive inspections would be impractical.

For large-scale releases, it is important to have easily recognized diagnostic features that allow the detection of failed confinement. In some cases, genetically based color traits, such as red kernels in corn, could be used to identify a particular transgene, assuming that the color trait stays tightly linked to the confined transgene. Distinctive phenotypes have been bred into some conventional crops, such as oilseed and "confectionary" sunflower, which have black seeds instead of striped seeds, respectively. Experimental lines of transgenic rice that have vitamin A precursors produce recognizable yellow grains, hence the name "golden rice" (Ye et al., 2000). An advantage of visually distinctive traits is that they are easy to identify with minimal expertise. However, a disadvantage is that they could be unreliable because of phenotypic plasticity, variable gene expression, or recombination that separates the genetic marker from the bioconfined transgene.

Transgenic methods could be used to introduce general or specific markers for the purpose of monitoring bioconfined transgenes. A general method could be to add a gene that expresses GFP, although that requires examining the plants in the dark with ultraviolet light—a technique with obvious limitations (Leffel et al., 1997). Another option is to assay for specific novel proteins in leaves or seeds using rapid enzyme-linked immuno-sorbent assays (ELISAs) that are similar to those at work in home test kits for pregnancy. Several companies market kits for detecting commonly used transgenes, such as antibiotic resistance proteins, that are often used as markers in genetically engineered plants. The kits are simple to use on leaf samples in the field, but false-negative results are common (Ilardi and Barba, 2001), and the cost of large-scale testing can be prohibitive.

In some cases, transgenic resistance to a particular herbicide could be inserted in the same construct as a bioconfined transgene to monitor for possible failure. Seed lots could be sampled and screened for the presence of rare, unexpected transgenes by applying the herbicide to large numbers of plants grown in field experiments (e.g., Scheffler et al., 1993). Herbicide-resistant survivors could be analyzed further to confirm the presence of the unwanted transgene. This method could be used on a case-by-case basis, but if the bioconfinement method failed it might lead to the unwanted spread of herbicide resistance as well as to the spread of the bioconfined transgene. However, in short-term, small-scale experiments, herbicide resistance could be a useful marker for testing the efficacy of new bioconfinement methods before they are used on a commercial scale.

In the future, unique DNA fingerprints could be linked to bioconfined transgenes to function as "bio-barcodes" ™ (Gressel, 2002). Those markers also could be useful for identifying nonconfined transgenes for labeling, but they require more elaborate and expensive laboratory techniques than are needed for the phenotypic traits mentioned above. Broothaerts and colleagues (2001) described a multiplex polymerase chain reaction (PCR) technique that simultaneously demonstrates the presence of a transgene sequence and an endogenous gene using a single reaction. Common transgene-specific primers were used in combination with conserved primers for polymorphic endogenous genes. The polymorphisms detected for the endogenous genes permitted the host plant's genotype to be determined, and they confirmed that the PCR had worked properly. The authors proposed the technology for use in protection against mislabeling of cultivars during subculturing and other laboratory and greenhouse operations, as well as for screening for transformants in the production of new transgene lines. The approach also would be useful in identifying cases of transgene escape into other cultivars or genotypes of the same species and their sexually compatible wild relatives.

Greater attention to the need for monitoring could lead to new and more effective approaches. For example, there is much interest in developing a "synthetic nose" remote sensing system that could identify portions of an agricultural field that are under attack by insects. This method would detect and profile volatile emissions from the plants (www.aginfo.psu.edu/News/march03/sentinel.html). Such devices are being developed for national defense and agronomic uses. Expression of transgenes for insect resistance also gives the genetically engineered plants a profile of volatile emissions that is different from that of wild-type plants of the same genotype, so it is possible that such transgene constructs could be detectable. Remote detection systems could be used to survey large natural areas for transgene or plant escapes at some point in the future, but that possibility is still quite speculative.

Given enough resources for statistically meaningful sampling efforts, it might be possible to detect failed bioconfinement, but there is still the problem of detecting failure early enough to mitigate or eradicate unwanted plants. If those plants reproduce and spread, either by further cultivation or by naturally occurring gene flow, subsequent efforts to stop the process could be futile. Therefore, plants that are judged to be serious enough risks should not be released because bioconfinement is always expected to be imperfect.

Population, Community, and Ecosystem Effects

Bioconfinement has rarely been used for cultivated plants, yet several new methods could become available within the next 5–10 years. Given the diversity of methods that are under development (Table 3-1), it is difficult to project environmental effects. Here, a few examples can be used to illustrate possible direct and indirect consequences of future bioconfinement strategies. Two types of effects are discussed: those in which the confinement method functions as intended, and those that result from an unintended breakdown.

For bioconfinement methods that rely on complete sterility, unwanted ecological or evolutionary effects are likely to be negligible *if* the method functions properly. For example, when a fully sterile crop or crop-wild hybrid produces no pollen, no viable seeds, and does not reproduce vegetatively, the transgene will not spread. Under what conditions could this pose a problem? A possible source of food for insects or wildlife could disappear if seed crops are eliminated through bioconfinement, although the ramifications could be relatively unimportant in some circumstances. For example, if vast tracts of planted, seed-producing trees, such as Douglas fir, were replaced with sterile trees, animal populations that depend on the seed source could be harmed. Whether that would threaten ecologically, economically, or socially important species would require further, case-by-case investigation.

Another hypothetical effect of transgenic sterility might occur if pollen from a crop with seed-specific sterility inundates small populations of wild relatives growing nearby. With extensive immigration of sterility-causing genes, the wild plants' seed production could be reduced (seeds sired by the transgenic pollen would be dead). Under some circumstances, this effect of "usurping" ovules and interfering with seed production might cause the native populations to shrink. However, few examples involving endangered wild relatives of crops have been identified (Hancock, 2003). Sexually compatible taxa that occur near crops often are weedy or colonizing species for which small population size is not a concern. If bioconfinement were indirectly responsible for greater contact between the crop and the wild relative, a possible case of unintended consequence could be argued. Moreover, if a crop's wild relatives are an important source of germplasm for further breeding, as is the case for perennial wild rice (*Oryza rufipogon*) in Southeast Asia (Lu et al., 2003), extra precautions might be needed to ensure that gene flow from a V-GURT does not exacerbate the erosion of valuable genetic diversity.

A more far-reaching fear among some members of the public is that sterility genes could spread throughout natural populations of wild relatives in a silenced (inactive) condition and later be reactivated, leading to massive die-off in populations of sexually compatible crop relatives. It is difficult to conceive of specific mechanisms that would support this hypothesis, but further study should be considered for transgenic sterility methods.

Other bioconfinement methods are intended to reduce the fitness of offspring from the crop or its crop-wild hybrids. Multiple scenarios for the fate of such fitness-decreasing transgenes should be considered to evaluate the effects of this process. First, if gene flow is extensive enough or the recipient population is small enough, deleterious transgenes could become fixed in feral or hybrid populations, perhaps leading to reduced populations. This type of "demographic swamping" could occur along contact zones between the crop and its wild relatives (e.g., Haygood et al., 2003). Lower fitness that is shared by all members of small populations along the contact zone could cause the population to shrink and perhaps disappear. A second and perhaps more likely scenario is that fitness-reducing transgenes would be purged by natural selection, a process that is likely to occur with many types of "domestication" crop genes that enter wild or weedy populations. Purging is expected to occur in populations for which gene flow is relatively low and the effective population size of wild relatives is larger than about 100 individuals. Large population size is common for most wild relatives of crop species.

Male sterility is a bioconfinement method that sometimes is misunderstood to be a danger to wild populations. Nontransgenic cytoplasmic male sterility has been used for decades to obtain hybrid seed in crops such as sunflower, canola, and sorghum (but not corn, for which mechanical detasseling is the commonly used method). Male sterility generally does not "breed true" or persist because of the large numbers of fertility-restoring genes that are found in cultivated and wild relatives of the crop (Besnard, 2000; Jan, 2000; Ohkawa, 1984; Yamagishi, 1998). In the future, new types of transgenic male sterility could come into common use for hybrid seed production in a wider variety of crops. Thus, male-sterile plants could be grown on much larger lands than at present, and it is possible that sterility would be passed on to plant offspring. If so, it is not expected that wild relatives of a crop would be harmed because fitness-reducing traits are quickly purged from large, interbreeding populations.

It is also important to consider the possible *indirect* effects of various bioconfinement methods. For example, how would a bioconfinement method affect populations of nontarget organisms, such as pollinators and other beneficial insects? Could the method harm animals at higher trophic levels in food webs because their prey are adversely affected? Could a novel trait like apomixis allow a vigorous cultivar to establish feral populations that invade natural areas? Also, would the method facilitate the cultivation of novel crops that produce unhealthy residues or facilitate environmentally damaging agricultural practices? How would those effects compare with existing problems caused by conventional agriculture? There is no reason to expect unwanted effects as a general feature of bioconfinement, but any large-scale release of novel GEOs should be accompanied by careful risk assessment. To thoroughly evaluate new methods it is necessary to examine anticipated benefits as well as possible risks of specific cases.

It also is useful to consider possible consequences when bioconfinement methods do not function properly, for example because of gene silencing or recombination that disconnects linked transgenes (Box 3-1). The ecological and evolutionary consequences of failed methods will depend on the characteristics of the transgenic plant, the environment in which it occurs, and the effectiveness of physical confinement. Failure of confinement methods—biological and otherwise—that are used to prevent pharmaceutical proteins in a commodity crop like maize from entering the food supply could lead to huge socioeconomic damage and unwanted effects on human health and nontarget organisms. Likewise, if bioconfinement fails to prevent the spread of an invasive horticultural variety, economic and environmental damage could be extensive. If bioconfinement is used with low-risk GEOs, however, the consequences of failure should be negligible. In general, the reason for investing in bioconfinement in the first place is usually strong enough to indicate the potential seriousness of the consequences of failure.

Specific consequences of bioconfinement failure will depend on the type and the scale of the damage, as is discussed in Chapter 2, reflecting the "hazard × exposure" equation used in academic discussions of risk assessment (see also Figure 2-1). In some cases exposure could be very small (e.g., Slavov et al., 2002, model on gene flow from poplar). However, in complex and constantly evolving ecological systems, the probability of exposure and the risk of harm from such exposure can be difficult to quantify empirically. Also, public perception of risk often is based on other, less tangible criteria. A basic tenet of this report is that bioconfinement is likely to fail to some extent, even when multiple methods are used to safeguard against failure.

4

Bioconfinement of Animals: Fish, Shellfish, and Insects

This chapter focuses on bioconfinement of two broad categories of genetically engineered organisms (GEOs): aquatic animals and insects. The aquatic animals considered are finfish (trout, catfish, tilapia) and shellfish, including mollusks (oysters, clams) and crustaceans (shrimp, crayfish). The Committee on the Biological Confinement of Genetically Engineered Organisms chose to focus on fish, shellfish, and insects because they are highly prone to establishing feral populations if they are intentionally introduced into the environment or if they escape from aquacultural or agricultural systems (NRC, 2002b). Captive lineages of those animals might serve as founders for genetically engineered lines, but they have undergone so little domestication that they often can reproduce and survive in suitable natural environments. Their reproductive and ecological traits are closely related to those of their wild relatives, thus raising the possibility of gene flow to or competition with wild relatives. Furthermore, many of the species of fish, shellfish, and insects targeted for genetic engineering have wild relatives in the environments they are likely to enter.

The chapter does not explicitly address bioconfinement of terrestrial livestock species because, as a group, they are less prone to becoming feral and causing ecological problems. There have been some important exceptions, however, such as feral goats in many countries and pigs, range chickens, and turkeys in several U.S. states (NRC, 2002b). When contemplating genetic engineering of livestock species that can become feral, it will be important to consider options for bioconfinement as part of the mix of feasible confinement methods. Some of the general approaches discussed in this chapter, such as the regulation of gene expression to prevent successful reproduction of escaped adults, also could be applied to livestock, although they would require tailoring to the biology of the species at issue. The committee's major findings and recommendations therefore apply generally to terrestrial livestock species that are prone to becoming feral.

Furthermore, this chapter does not directly address bioconfinement of laboratory research animals, such as inbred strains of mice or rats. Laboratory animal strains typically are held in rearing or research facilities with multiple physical containment features and high security against theft. If research with transgenic lines of laboratory animals were to rely more heavily on bioconfinement than on physical confinement, the committee's findings and recommendations also would apply generally to those species (cat, mink) that might escape the laboratory and become feral in an accessible ecosystem.

Biotechnologists are developing transgenic fish and shellfish for a diversity of purposes (Table 2-2; Kapuscinski, 2003, and references therein). The proposed application of many transgenic lines is in aquaculture to produce human food, and it focuses on increasing growth rates and food conversion efficiency or improving disease resistance. Scientists also are developing transgenic lines for use as biofactories to produce pharmaceuticals, industrial chemicals, or dietary supplements; in bioremediation to remove contaminants from water; as water quality sentinels to detect contaminants that damage the genes of living organisms; and for biological control of nuisance aquatic species. Some degree of mechanical and physical confinement is possible for some of the proposed transgenic fish and shellfish (Scientists' Working Group on Biosafety, 1998). In other cases transgenic lines will be introduced into natural waters, either deliberately as in biological control applications, or unintentionally by escape from floating net cages, outdoor ponds in flood-prone zones, and flow-through raceways. One also can envision proposals to deliberately release hatchery-propagated fish or shellfish, such as cold-tolerant or endemic-pathogen-resistant lines, to establish a new sport or commercial fishery, or to augment an existing fishery.

There are several reasons for developing genetically engineered insects. Agricultural applications include transgenic-based sterile males (replacing radiation-induced sterile males) for mass releases in biological control of pest insects, visual transgenic marking with markers such as green fluorescent protein that can be used to evaluate effectiveness of sterile insect releases, and genetic engineering of beneficial insects (predators and parasitoids of pest insects) for resistance to insecticides to allow simultaneous use of both methods of controlling pest insects (Braig and Yan, 2002; NRC, 2002b; Wimmer, 2003). Genetic engineering also has been proposed for introducing disease resistance and other desirable traits into domesticated insects, such as honeybees and silkworms, and to turn insects into biological factories for mass production of valuable proteins such as collagen, which could be produced by silkworms (Tamita et al., 2003). Finally, genetic engineering research is under way to disrupt transmission of diseases by mosquitoes and other vectors (Braig and Yan, 2002; NRC, 2002b; Spielman et al., 2002). In this last application, bioconfinement is not an option because achievement of disease suppression requires that the released transgenic insects mate widely with wild-types to spread their transgenes throughout the population.

The discussion in this chapter assumes that the transgenic animals are dioecious—male and female reproductive organs are in separate individuals and each individual is of one sex throughout its lifetime. However, non-dioecious modes of reproduction, such as hermaphroditism and partheno-genesis, occur in some species of fish, mollusks, and crustaceans, some of them aquaculturally important (reviewed in Appendix B of ABRAC, 1995). Bioconfinement methods discussed in this chapter that target sexual repro-duction could fail to achieve the desired amount of confinement or, in some cases, could simply be infeasible in hermaphroditic and parthenogenetic species.

Hermaphroditic individuals have male and female organs; parthenogens have some form of clonal inheritance of genomes (Moore, 1984). Hermaphro-dites occur in some species of sea bream (Buxton and Garrett, 1990), a family of finfish with several species produced in aquaculture, and at least one species that is already the subject of gene transfer for growth enhance-ment (Zhang et al., 1998) and reported to exhibit hermaphroditism (Huang et al., 1974). Parthenogenesis occurs in strains of aquacultural crustaceans such as *Artemia* (brine shrimp) (Triantaphyllidis et al., 1993) and *Daphnia* spp. (Hebert et al., 1993). Self-fertilizing hermaphrodites and true parthenogens, which do not require the physical stimulus of sperm to induce embryogenesis, pose the greatest challenge for confinement because the escape of just one fertile individual could result in the establishment of an entire population.

BIOCONFINEMENT OF FISH AND SHELLFISH

Bioconfinement methods currently in practice for fish and shellfish either reduce the spread of transgenes and transgenic traits through disrup-tion of sexual reproduction or rely on ecological characteristics of the production site that are lethal to some life stage of an escaping organism.

Disruption of Sexual Reproduction

Methods for disruption of sexual reproduction include induction of triploidy or interploid triploidy—causing embryos that normally bear two sets of chromosomes to carry a third set; induction of monosex lines; and crossing two closely related species to produce viable but infertile hybrids (sterile, interspecific hybrids). The methods sometimes are combined, particularly triploidy monosex production.

Sterilization through Induction of Triploidy

Triploidy induction involves application of hydrostatic pressure or temperature or chemical shock at the appropriate number of minutes after egg fertilization to disrupt the egg's normal extrusion of a polar body that contains a haploid set of chromosomes. The resulting retention of the polar body leads to an embryo that bears a pair of haploid chromosome sets from the female (instead of the normal single set) and a third set from the male (Figure 4-1). The presence of the odd set of chromosomes presumably causes mechanical problems involving the pairing of homologous chromosomes during each cell division (Benfey, 1999), and this disrupts the normal development of gametes to some extent, as explained below.

Triploidization is much better developed for finfish and mollusks than it is for crustaceans produced in aquaculture. Protocols for large-scale induction of triploidy have been worked out for a number of commercially important fish and mollusks, including various trout and salmon species, channel catfish, African catfish, various tilapia species, various carp species, oysters, and clams (reviewed in Beaumont and Fairbrother, 1991; Benfey, 1999; Li et al., 2003; Tave, 1993; Thorgaard, 1995). However, protocols need to be developed and optimized for each species. Induction of triploidy in crustaceans might be possible only in shrimp species that spawn free eggs (genera *Litopenaeus* and *Penaeus*) and not in those species, such as freshwater prawns (*Macrobrachium rosenbergii*), whose females incubate their fertilized eggs (Beaumont and Fairbrother, 1991; Dumas and Campos Ramos, 1999). Researchers are in the early stages of developing reliable protocols for triploid induction in marine shrimp, but recent efforts are part of the increased interest in the genetic improvement of shrimp—from traditional breeding to gene transfer (e.g., Dumas and Campos Ramos, 1999; Fast and Menasveta, 2000; Li et al., 2003).

Strengths

Triploidy induction has become widely accepted as the most effective method today for producing sterile fish for aquaculture (Benfey, 1999; Tave, 1993). It is the best-developed method of disrupting sexual reproduction, and it has the most complete scientific documentation of strengths and weaknesses. Triploidy has been used on commercial rainbow trout and Atlantic salmon farms (Donaldson and Devlin, 1996). Triploid Pacific

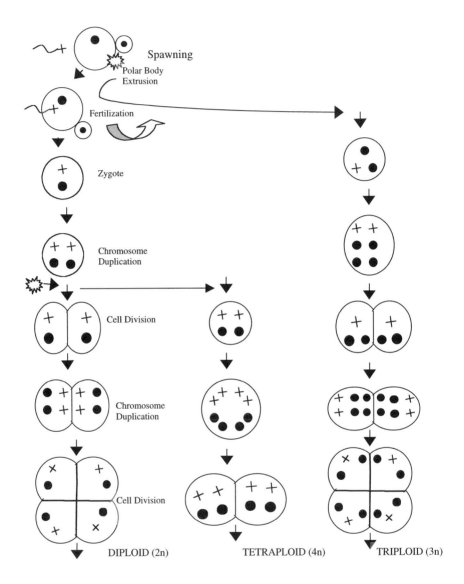

FIGURE 4-1 Normal steps in gamete fertilization and early cell division that lead to the development of a normal diploid ($2n$) fish or shellfish embryo. Induction of triploidy ($3n$) or tetraploidy ($4n$) occurs by temperature shock, chemical shock, or pressure at an appropriate time after fertilization:

✿ denotes the point at which the shock is applied;

● denotes one haploid chromosome set derived from the female parent; and

+ ● denotes one haploid chromosome set derived from the male.

SOURCE: Adapted from Donaldson, unpublished data.

oysters make up 30% of all Pacific oysters farmed on the West Coast of North America (Nell, 2002), not so much for bioconfinement as to prevent yield losses associated with sexual maturation in production animals. Procedures for inducing triploidy are easy to learn and require relatively inexpensive, simple equipment. It is feasible to screen individuals nonlethally and to collect blood, hemolymph (the shellfish equivalent of blood), or another small tissue sample, for the presence or absence of the triploid condition (Harrell and Van Heukelem, 1998; Nell, 2002; Wattendorf, 1986). Individual screening has long been required for large-scale stocking of triploid grass carp in Florida (Griffin, 1991; Wattendorf and Phillippy, 1996). Farmers interested in stocking this alien species into their irrigation canals to help control aquatic nuisance weeds are required to have each fish tested and certified as triploid before release.

Weaknesses

The incomplete success in producing triploids is a major problem, particularly for treating large batches of newly fertilized eggs. Several limitations to screening and detection affect success with culling individuals that fail to become triploid. The degree of functional sterility in triploids varies, depending on the species and sex of the fish. A small percentage of mosaic individuals (bearing a mix of diploid and triploid cells) also can compromise sterility if their gonads are diploid and thus develop into normal, fertile gametes. Sterile individuals that still enter into courtship behavior could disrupt successful reproduction of wild relatives, and recurring large escapes of sterile individuals could heighten competition with or predation on wild species. Commercial aquaculturists could resist adopting sterile lines of fish and shellfish. These weaknesses and possible mitigation are explained more fully below.

Variable atriploidy: The percentage of triploids produced from a treated batch of eggs varies greatly by species and strain, method, pretreatment water temperature (when induction is by heat shock), and egg quality (see review in Galbreath and Samples, 2000). Reported success rates in finfish range from 10% to 100% (Galbreath and Samples, 2000; Johnstone et al., 1989; Maclean and Laight, 2000). Although little has been published about large-scale treatments, Johnstone and colleagues (1989) reported 100% triploid fish with a 90% survival rate relative to controls in a large-scale trial involving pressure shock on 50,000 eggs per hour. Commercial aquaculture companies that produce and market triploid fish are likely to have closely held data on success rates of large-scale pressure shock and temperature shock treatments for triploid induction. Effectiveness in shellfish ranges from 85% to 95% in oysters (personal communication, S. Allen, School of Marine Science, Virginia Institute of Marine Science, Gloucester Point,

2003) and from 63% to 100% on application of the "optimum" protocol in one shrimp species (Li et al., 2003).

Screening to mitigate failed triploidization: Less than total triploid induction can be mitigated by screening treated individuals and then removing the nontriploids before they are transferred from hatcheries to much less secure grow-out facilities, such as outdoor ponds or open-water cages (Kapuscinski, 2001; Kapuscinski and Brister, 2001). Mass screening is feasible through particle analysis or flow cytometry; particle analysis allows almost instantaneous results (Harrell and Van Heukelem, 1998; Nell, 2002; Wattendorf, 1986). Both methods permit non-lethal screening of larger juvenile life stages because they require only minute quantities of blood (as little as 1 μL or one drop) or disaggregated tissue (Harrell and Van Heukelem, 1998). Detection limits and operator error are facts of life for either method. The critical management issue regarding the amount and verifiability of the bioconfinement provided by induction of triploidy is whether to screen every individual destined for grow-out or only a sample of each production lot. Such a decision should consider the risk, severity of consequence (Table 2-1 and Figure 2-1, Chapter 2), and the extent to which adequate additional confinement measures are in place. The discussion of transgenic salmon presented in Box 4-1 illustrates the point.

Use of Tetraploids to Maximize Triploid Percentage

The failure rate in producing triploid individuals can be reduced or avoided altogether by making triploid individuals via crosses between a tetraploid adult (usually the female) and a diploid adult (Guo et al., 1996; Tave, 1993; Xiang et al., 1993). Newly fertilized embryos are induced to become tetraploid (bearing four sets of chromosomes instead of the normal two sets) in the first generation (Figure 4-1). Then the diploid eggs produced by a tetraploid female are crossed with the normal haploid sperm of a male to generate all-triploid offspring in the next generation. The offspring are called interploid triploids, or "genetic" triploids, to distinguish them from induced triploids. The generally poor survival and performance of tetraploid fish (Donaldson and Devlin, 1996), however, prevents large numbers of individuals from reaching sexual maturity. This has discouraged large-scale production of interploid triploids in finfish and could be an obstacle for bioconfinement of genetically engineered species. Much better performance of tetraploids has been reported in oysters produced by crossing eggs from triploid females with sperm from diploids (Allen and Guo, 1998). Most important for bioconfinement, the yield of interploid triploid oysters can be very high; one researcher has reported that 99.3% of more that 2,100 offspring were triploid (S. Allen, unpublished data), and the approach is in use by some commercial oyster farms (Nell, 2002). The

BOX 4-1
Proposed Bioconfinement of Transgenic Atlantic Salmon

Aqua Bounty Farms, a biotechnology company, has applied to the U.S. Food and Drug Administration (FDA) for commercial approval of transgenic, growth-enhanced Atlantic salmon (Office of Science and Technology Policy and Council on Environmental Quality, 2001). The company intends to sell transgenic embryos or newly hatched fry to industrial salmon farms. The salmon farms would raise the juvenile fish in confined hatchery systems, usually consisting of land-based tanks and ponds, and then transfer the older smolts (a life stage that can thrive in seawater) to less confined, floating cages in coastal marine waters. This has raised concerns about potential ecological harm, particularly to already severely depleted populations of wild Atlantic salmon. Introduction of a new threat to wild Atlantic salmon would occur in the face of costly and complicated efforts under way to recover declining Atlantic salmon populations (e.g., NRC, 2002d, 2004).

Within the native range of Atlantic salmon, the primary ecological concern is whether the movement of transgenes into wild populations has a higher, equal, or lower potential to depress fitness (Kapuscinski and Brister, 2001; NRC, 2002b; Pew Initiative on Food and Biotechnology, 2003). Computer simulations have suggested scenarios involving earlier age at sexual maturity or larger size of reproducing adults—traits often associated with faster growth rates in fish—combined with moderately lower, equal, or higher viability in transgenic salmon than in wild fish, that could pose a heightened threat to the fitness of wild populations (Muir and Howard, 2001; NRC, 2002b; Pew Initiative on Food and Biotechnology, 2003). It is unclear whether the company has collected the data needed to assess whether its transgenic salmon fit any of these scenarios, partly because such data have not been reported in scientific journals and partly because of the lack of transparency in the FDA drug approval process (Kapuscinski, 2001; NRC, 2002b; Pew Initiative on Food and Biotechnology, 2003).

In salmon farming regions outside the natural range of Atlantic salmon (e.g., Chile, New Zealand), the main question would be whether the net fitness of transgenic salmon is higher or lower than in currently farmed strains and thus whether the transgenic fish would be more or less of a threat to invade native regions (NRC, 2002b; Pew Initiative on Food and Biotechnology, 2003). Heightened invasiveness could pose a risk to native fish and other aquatic species through predation or competition (Scientists' Working Group on Biosafety, 1998).

The basis for concern is the increasing documentation of thousands to hundreds of thousands of farmed salmon that escape from cages that have been damaged by storms, predators, or wear and tear (e.g., Carr et al., 1997; Gross, 1998, 2001; Thomson, 1999). Most escapees are smolts, postsmolts, and adults; all of which can move from one habitat to another and interact directly or indirectly with wild salmon (NRC, 2002d, 2004). As they mature, escapees have been found to migrate into rivers (Hansen and Jonsson, 1991; Whoriskey and Carr, 2001; Youngson et al., 1997) and to spawn in those rivers (e.g. Clifford et al., 1998; Lura and Seagrov, 1991; Webb et al., 1991). Breeding between farmed salmon escapees and wild salmon can depress the reproductive success and competitive ability of wild populations through various mechanisms during the breeding season and in the next generation (NRC, 2002d, 2004).

continued

BOX 4-1 Continued

A less-examined exposure route that could be significant (Stokesbury and LaCroix, 1997) is the escape of juvenile salmon from freshwater hatcheries operated by salmon-farming companies (NRC, 2002d). Competitive interactions between farmed and wild salmon juveniles for food and space in rivers can lead to displacement of wild fish and to depressed productivity of the wild population (Fleming et al., 2000; McGinnity et al., 1997).

To reduce the likelihood of damage, Aqua Bounty Farms has suggested that it will sell nothing other than batches of embryos or newly hatched fry that are all-female and subjected to mass-scale induction of triploidy. That combination takes advantage of the fact that triploid salmon females cannot produce viable eggs even though triploid males can still produce viable sperm (reviewed above in this chapter). Resources for achieving strict confinement can focus on holding the transgenic broodstock needed to propagate the all-female progeny in one or a few facilities. The proposal also would protect the company's patent on the marketed line of transgenic fish by preventing salmon farmers from propagating the line because they would be required to purchase production fish for each grow-out cycle.

The Aqua Bounty Farms proposal has two important weaknesses. First, it depends heavily on screening to identify and cull failures of triploid induction. The critical management issue is whether to screen every individual prior to transfer to grow-out facilities or only a sub-sample of each production lot, as discussed above in this chapter. Such a decision should consider the level of risk and severity of consequences (Table 2-1 and Figure 2-1, Chapter 2) and adequacy of the integrated confinement system (Chapter 6). The net fitness method (Muir and Howard 2001, 2002) provides a means to estimate—in a secure setting—the probability of spread of the transgenes if fertile transgenic salmon were to escape, although it cannot predict the severity of the harmful consequence from such transgene spread. This estimate would help decision makers determine whether to screen all or only a sub-sample of each production lot. If they choose sub-sampling, this estimate would help determine the appropriate sample size as a function of the predicted severity of harm, the probability of harm given an escape of fertile salmon has occurred, and the probability of escape of fertile fish.

Individual screening followed by culling of diploids would be the more prudent choice for farming all-female, triploid transgenic Atlantic salmon in open-water cages in areas—such as the Maine coast—where wild populations are already depleted severely. Eight populations in Maine are listed as endangered under the terms of the Endangered Species Act (NRC, 2002d). Fewer than 100 sexually mature adults returned to these eight rivers in 2000–2002 (NRC, 2002d), and fish traps placed on three of the rivers intercepted up to 65 farmed salmon escapees each year (1993–2001). That number represents a range of 0–100% of returning adults (NRC, 2002d). Those data suggest that even a small number of escaped fertile transgenic fish could constitute a major cohort of interbreeding adult fish in Maine's rivers.

continued

BOX 4-1 Continued

The use of marine fish cages that are suspended in coastal waters makes it nearly impossible to meet the committee's recommendation to institute integrated confinement systems (Chapter 6). The cages provide weak physical confinement and preclude "end of the pipe" confinement measures such as imposing lethal temperatures or chemical treatment of effluent water through which fish might escape. Thus, the confinement system relies heavily on the biological dimension and hinges specifically on the triploidization success rate. It also depends on the statistical power of detecting fertile diploid fish at different frequencies and sample sizes if culling relies on screening a sample rather than each fish in the lot.

A conservative estimate indicates that the cost of screening individual salmon by flow cytometry would add $0.02 to $0.04 per 1 kg of fish to the market cost of farmed Atlantic or chinook salmon (Kapuscinski, 2001). The estimate is considered conservative because it is based on small-scale tests (Wattendorf, 1986), and it does not account for the economies of scale afforded by the use of flow cytometry screening (Harrell and Van Heukelem, 1998). It also does not include the reduced price of labor or the time saved that could be achieved through computer automation techniques. In any event, the cost of individual screening is a fraction of the current market price of salmon molts, trout fingerlings, or other early-life stages purchased by grow-out farmers.

The second weakness of the proposed bioconfinement is the potential for reproductive interference or other competitive interactions caused by periodic large escapes and possible migration of all-female triploid salmon into rivers. Reproductive interference would occur if the females had reproductive hormone concentrations sufficient to cause them to ascend rivers, mate with wild males, and produce infertile broods. Given that males can spawn with more than one female, this would be of greatest concern where most available females were sterile-farm escapees, because the total number of wild adults that return to the rivers would be extremely low in succeeding generations. A lack of appropriate research on the courtship and migratory behavior of triploid all-female salmon makes it difficult to assess the extent to which reproductive interference is a concern.

The weaknesses of the proposed bioconfinement measures could be avoided by combining bioconfinement with the much more reliable physical confinement afforded by farming salmon in land-based facilities, ideally in closed-loop recirculating aquaculture systems (Kapuscinski, 2003). The salmon farming industry is under increasing pressure to solve a host of environmental problems posed by cage farming regardless of the possible adoption of transgenic salmon. A few entrepreneurs have responded by establishing land-based salmon farms in North America. The initial capital costs and the higher operating costs of land-based operations are major disincentives to an industrywide switch from sea cage to land-based production systems. However, Aqua Bounty Farms has publicly suggested that the cost advantage of producing faster-growing transgenic salmon could give salmon farming companies enough economic leeway to make the switch to land-based production (McClure, 2002).

production of second-generation tetraploid Pacific oysters (Guo et al., 1996) is stimulating work to establish tetraploid breeding lines that will remove the need to continuously induce tetraploidy.

Mosaic individuals: A small percentage of putative triploids can become mosaic—bearing some diploid and some triploid cells—as has been found in studies of fish and oysters (Benfey, 1999; Harrell and Van Heukelem, 1998; Hawkins et al., 1998). Bioconfinement would be compromised if cells within gonadal tissue were mosaic, but no published data were found on searches for this in fish. Research in Pacific oysters has shown that some triploids revert progressively over their lifetime to a mosaic state, raising the possibility that they could produce viable gametes (Calvo et al., 2001; Zhou, 2002). Reductions in triploidy have ranged from 2% to 10% to more than 20% in Pacific oysters (Allen et al., 1996; Nell, 2002). One researcher reported reversion to mosaics to be an order of magnitude lower in interploid triploid oysters (0.6%) than in induced triploids (2.5%–10%), although both types had a low incidence of "streakers" that revert to diploidy in all or nearly all tissues (Standish Allen, unpublished data).

Variable functional sterility: Even when the induction is successful, the amount of functional sterility achieved is highly variable. Triploidy in finfish disrupts gonadal development somewhat in males but more fully in females, with some exceptions (Thorgaard and Allen, 1992). Where triploid females fail to produce viable eggs, combining triploidy with production of all-female lines substantially increases the effectiveness of bioconfinement.

Disrupting reproduction in wild relatives: Triploid sterilization would not completely remove the need to assess the ecological consequences of escaped GEOs because triploids of some species have enough sex hormones to cause them to engage in normal courtship and spawning behavior. Escaping triploid fish could interfere with the reproduction of wild relatives by mating with fertile wild adults, leading to losses of entire broods and lowering of reproductive success. The most severe consequence would be reproductive interference in already declining, threatened, or endangered species. Nearly all U.S. salmon populations other than those in Alaska are at risk. There has been little research to investigate the extent to which triploid adults of different fish species retain normal reproductive behavior. In trout and salmon, the concern appears to be mostly with triploid males (Cotter et al., 2000; Inada and Taniguchi, 1991; Kitamura et al., 1991). The risk could be lessened through production of transgenic lines of sterile females (Donaldson and Devlin, 1996). In one of the few field tests of the behavior of triploid fish released into the natural environment, triploid adult Atlantic salmon migrated back from the ocean to natal freshwaters at a much lower rate than did control salmon, thus reducing the population that could attempt to mate with wild fish (Cotter et al., 2000). Virtually nothing is known about the extent to which triploid shrimp and other crustaceans retain normal reproductive behavior.

Heightened competition or predation: It also would be necessary to assess possible ecological disruptions if large numbers of triploid transgenic individuals were to enter the environment on a *recurring* basis, either through escape from aquaculture operations or through intentional introductions to support a fishery (ABRAC, 1995; Kapuscinski and Brister, 2001). Sufficient numbers of sterile transgenic adults could survive and grow for an indeterminate period beyond the normal lifespan, given that they did not expend energy on reproduction, and those fish could heighten competition with wild relatives or prey on other species (Kitchell and Hewitt, 1987). This concern cannot be dismissed easily, given the high frequency and large number of fish that escape from some commercial aquaculture operations. For instance, there have been large recurring escapes of farmed salmon in coastal waters with heavy concentrations of floating farm cages (e.g., Carr et al., 1997; Gross, 1998, 2001; Thomson, 1999).

Farmer reluctance: Finally, aquaculture producers could be reluctant to adopt sterile lines of transgenic fish or shellfish for two reasons. First, in some species, current sterilization methods can depress survival or growth or exacerbate morphological deformity (Benfey, 1999; Wang et al., 1998), thus offsetting the advantages of any transgenically enhanced traits. Triploidization was shown to depress growth enhancement by as much as 41% in transgenic tilapia that bore a growth-promoting gene construct (Razak et al., 1999). Results from a study of autotransgenic mudloach suggest that combining a different, more effective growth-promoting construct with triploid induction could eliminate that drawback (Nam et al., 2001a, b). The growth acceleration of the sterile (triploid) autotransgenic mud loach was 22–25 times higher than that of nontransgenic diploids. This represented a relatively modest decline compared with the more than 30-fold growth acceleration of the fertile (diploid) autotransgenic fish. Triploid oysters and other mollusks generally grow more and faster than do their diploid counterparts (Guo, 1999). Little is known about effects of triploidy on the overall performance of crustaceans (Fast and Menasveta, 2000), although in one laboratory study triploids of a marine shrimp species grew 30% larger than did diploids (Xiang et al., 1999).

Some aquaculture producers also are reluctant to purchase new batches of sterile fish eggs or fry every growing season rather than growing out fish reproduced from their own broodstocks. Each time, they must pay a patent royalty for the fish (Kelso, 2003). This becomes a concern under the likely scenario that biotechnology companies will sell only sterile transgenic fish eggs or fry to farmers.

Combining Triploid Sterilization with All-Female Lines

Triploidy is sometimes combined with production of all-female (monosex) lines if triploidy alone disrupts gonadal development somewhat in males and more in females. The problem has been documented in triploid lines of several commercially important finfish species—trout, salmon, grass carp, and tilapia (Liu et al., 2001). Ovarian growth is typically greatly retarded, whereas testes grow to near normal size, so that triploid males often produce small amounts of viable sperm that have aneuploid chromosome numbers and other abnormalities. In most species, fertilization of eggs with this viable sperm from triploid males produces progeny that die as embryos or larvae. Typically, the triploid females do not produce mature oocytes, although several studies that went beyond the normal first time of sexual maturation in diploids did report occasional production of mature oocytes by triploid females (Benfey, 1999). As technologists seek to induce triploidy in more species, it will be important to test for the extent of sterilization achieved in both sexes. For now, the production of all-female lines of triploids in fish and shellfish is the best way to maximize disruption of gonadal development in both sexes (Benfey, 1999; Donaldson and Devlin, 1996). Commercial farming of all-female lines or all-female and triploid lines is now widespread for several species of salmon and trout in North America, Europe, Asia, and Tasmania (Donaldson and Devlin, 1996). Attempts to produce monosex lines of shrimp, however, have yet to be successful (Moss et al., 2003).

A few studies of triploid induction, in common carp and channel catfish, reported sterility in males and females (Gervai et al., 1980; Wolters et al., 1982). If such results are repeatable, then induction of triploidy alone—without the additional production of all-female lines—could be an adequate method of sterilization. Note, however, that production of all-female lines alone—without triploid sterilization—is not an adequate method of bioconfinement, particularly if the species produced in an aquaculture system has wild relatives in accessible ecosystems.

The methods for production of all-female lines of fish vary, depending on whether the species has an XX/XY sex-determining system or a WZ/ZZ sex-determining system. They are well described and have been used successfully in a variety of aquacultural species (reviewed by Tave, 1993). The production cycle for integrating triploidy induction into an all-female line has been developed for salmon, trout, and other species with an XY sex-determining system (Donaldson and Devlin, 1996; Figure 4-2). Applying this production cycle to transgenic fish involves developing an all-female line of transgenic fish, then fertilizing transgenic eggs with milt from the sex-reversed females, and inducing triploidy on the newly fertilized eggs.

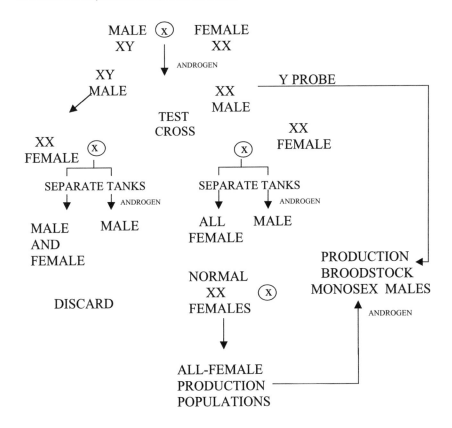

FIGURE 4-2 Production cycle for all-female lines of fish in species with an XY sex determination system.
SOURCE: Adapted from Donaldson and Devlin (1996).

Triploidy induction must occur every time the all-female transgenic line is bred to produce offspring for grow-out.

Strengths

The combination of all-female lines with triploidy circumvents the problem of incomplete sterilization in triploid males. Protocols are well established in some commercially important species and should be fairly easy to develop for others.

Weaknesses

Two or more generations are needed initially to establish an all-female line, with the exact number depending on the protocol used and the sex determination system of the species. Applying this approach to confine transgenic lines or even nonindigenous species requires maintenance and propagation of all-female broodstocks under strict confinement; this adds extra cost to commercial development.

Production Site Characteristics

It could be possible to confine fish and shellfish to some extent by producing them in aquaculture facilities where one or more of the following criteria are met:

• The facility is located in an arid region with interior drainage only and no permanent water bodies, such that all surface water runoff either percolates into the ground or evaporates (ABRAC, 1995).
• All accessible ecosystems lack wild relatives, thus precluding gene flow from any fertile genetically engineered organisms into wild populations.

It is important to consider wild relatives not only of the same species as the GEO but also of closely related species. Many fish and shellfish can hybridize with closely related species. But meeting this condition of isolation alone would not prevent establishment of a freely reproducing population of GEOs, essentially as a new invasive species, in one or more accessible ecosystems.

• In all accessible ecosystems, water chemistry—temperature, salinity, pH, or concentrations of specific constituents—is proven to be lethal to one or more life stages of the genetically engineered line.

Care should be taken in applying this approach because predictions of lethal conditions for fish populations have sometimes been wrong, as exemplified by the pink salmon invasion of the Laurentian Great Lakes. The long-held assumption was that subadult and adult life stages could not survive in freshwater habitats (reviewed by Kapuscinski and Hallerman, 1991). Conditions that are lethal for the unmodified parental organism might not accurately predict the response of a transgenic line. Each line should be tested directly for lethal water chemistry conditions because the inserted genes could alter physiological traits that influence the animals' response to water chemistry. Growth hormone, for instance, affects fish growth and salinity tolerance. It would be important to determine seasonal and annual variations in water chemistry: Favorable conditions can occur periodically but persist long enough to allow initial establishment of transgenic individuals, followed by natural selection on their offspring for adaptation to more typical conditions. Adaptive evolution to new environments can happen surprisingly rapidly in fish populations: Guppies introduced to a new wild stream environment showed adaptive evolution in only seven generations, a mere four years for that species (Reznick et al., 1997).

Strengths

This approach does not require any additional manipulations of production organisms or cost associated with applying and verifying manipulation. The degree of successful confinement should remain relatively constant, except for natural variations in environmental conditions (such as warming of water temperatures) that could make the natural biological barrier ineffective.

Weaknesses

The cost of achieving optimum growing conditions within a commercial fish or shellfish farm is likely to increase with the inhospitability of the surrounding ecosystem to escapees. For instance, locating a facility in an arid environment is likely to increase the cost of obtaining and maintaining a sufficient water supply. Uncertainty tends to be high regarding whether the cumulative ecological conditions will prevent reproduction or survival of escapees, particularly because important ecological factors often change over time. Proposals to use site characteristics, such as those listed above, as the main form of bioconfinement, should undergo considerable scrutiny by an interdisciplinary team with expertise in the broad relevant principles and in site-specific aspects of climatology, animal ecology, community ecology, hydrology, and watershed science. Additional areas of expertise and local knowledge could be necessary, depending on the case.

Gene Blocking and Gene Knockout

The growing base of information about the function of specific animal genes, inducible promoters, and ways of blocking gene expression or of completely knocking out target genes could be harnessed to disrupt reproduction or survival of escaping fish and shellfish. It also could be applied to biologically confine insects (discussed later in this chapter). Applying this broad approach to bioconfinement of GEOs would require adding another engineered-DNA construct whose role is to disrupt some essential life-enabling function in escaped animals. A variety of possible approaches are in the early stages of research and development. The approaches discussed below include insertion of inducible transgene constructs that block or disrupt expression of an essential endogenous gene only after animals escape captivity; RNA interference (RNAi) genes that block or misexpress translation of an essential endogenous gene, combined with exogenous administration of the affected essential compound to animals in captivity; externally administered gene-specific substances that interrupt expression of a gene essential for development of normal reproductive organs; and gene knockout. Some of the research focuses on zebrafish, a model research organism for elucidating the workings of genes in vertebrate development (e.g., Zhao et al., 2001). Methods developed for zebrafish could be applied to bioconfinement of other transgenic animals. Other research has aimed, from the outset, at developing bioconfinement techniques for transgenic fish and shellfish (e.g., Thresher et al., 1999). Table 4-1 lists some examples of genes that have been identified in fish and that could be targets for genetic engineering to disrupt reproduction, survival, or an essential process in development.

Inducible Transgenic Gene Blocking or Misexpression

This approach involves inserting a construct designed to block expression of an endogenous gene that is essential in the development of viable gametes or embryos. The construct includes a sequence for a blocker molecule that prevents expression (or at least causes misexpression) of an endogenous gene. Expression of the blocker is controlled by an inducible promoter, ideally one that is triggered by the presence or absence of a compound that can be added to the food of captive animals. Other parts of the construct allow reversible activation and repression of the inducible promoter and hence of the expression of the blocker.

This general approach is illustrated by a patent for repressible sterility in animals, including fish and oysters—the "sterile feral" technology (Thresher et al., 1999). In one example described in the patent, the aim is to block the zebrafish's endogenous gene for bone morphogenetic protein (*zBMP 2*). This protein is expressed only during early larval development and is essential for normal development of specific tissues, such as blood. Blocking gene expression is therefore lethal to zebrafish embryos. To produce a blocker molecule, the transgenic construct has a sequence that encodes a form of RNA, such as double-stranded RNA (dsRNA), which binds to the endogenous *zBMP 2* gene and prevents its expression. The promoter that drives the expression of the blocking RNA normally activates only during embryogenesis; when this happens, the zebrafish embryos die. However, to allow normal embryonic development in captivity, the construct also contains DNA sequences that respond to the presence or

TABLE 4-1 Genetic Bioconfinement Strategies for Fish

Gene	Bioconfinement Strategy	Reference
Aromatase	Block to produce all-male line	Nowak, 2002; Genbank, cited by Donaldson and Devlin, 1996
Estrogen receptor	Sterilization	Genbank, cited by Donaldson and Devlin, 1996
Gonadotropin-releasing hormone	Ablate maturation	Genbank, cited by Donaldson and Devlin, 1996
Gonadotropin subunits	Ablate maturation	Genbank, cited by Donaldson and Devlin, 1996
Protamine	Sterilize males	Genbank, cited by Donaldson and Devlin, 1996
Steroid 17-α mono-oxygenase	Block steroidogenesis	Genbank, cited by Donaldson and Devlin, 1996
Vitellogenin	Sterilize females	Genbank cited by Donaldson and Devlin, 1996
Zebrafish, bone morphogenetic protein	Disrupt embryonic development ("sterile feral" technology)	Thresher et al., 1999

NOTE: Genes that regulate a step in reproductive development and potential harnessing for bioconfinement are cloned from salmon or trout species, unless stated otherwise.
SOURCE: Adapted from Table 2, Donaldson and Devlin, 1996.

absence of a "repressor" molecule. In this case, the repressor is the antibiotic tetracycline or analogues such as doxycycline (Gossen and Bujard, 1992; Gossen et al., 1995; Kistner et al., 1996). Addition of the antibiotic to the water or food supply represses expression of the blocking dsRNA, thus allowing normal expression of *zBMP 2* and normal embryogenesis.

Strengths

The ability to repress gene blockage allows normal performance of animals while they are in captivity. This method also allows for building in multiple redundancy by stacking sequences to block expression of different essential genes and at different stages of development. This could significantly reduce the failure rate of bioconfinement. However, there probably are limits to and complications with stacking genes in transgenic animals.

For instance, genomic integration of large stacked-gene constructs could increase disruption of favorable endogenous genes, thus imposing a practical limit to the degree of redundancy achievable by this method alone.

Weaknesses

The blockage of expression of the targeted gene might never reach 100%, raising problems similar to those regarding success rates of triploid induction. Data on effectiveness await completion of development of transgenic fish lines with stable inheritance and expression of a sterile feral genetic construct (R. Thresher, CSIRO Marine Research, personal communication, May 20, 2003). The blocking of embryonic viability was still well below 100% in early experiments in which the dsRNA was simply injected into whole embryos (Thresher et al., 1999), indicating the need for considerable research before this technology can be used for commercial bioconfinement. The expression of the blocker molecule or its promoter could be turned off by methylation (NRC, 2002b) or breakup of the construct during recombination or mutation. This should occur only in a very small fraction of fish that escape from an aquaculture operation or of their fertilized embryos, assuming that biotechnologists would have confirmed stable integration, transmission, and expression of the sterile feral construct before commercialization. Natural selection, however, would strongly favor individuals in which the sterility genes failed to express; even a small failure rate among escapees could multiply fairly quickly into a large incidence of fertile transgenic individuals in the wild, especially in cases where the main transgene (not the sterile feral construct) confers some selective advantage over that of untransformed conspecifics.

Another potential cause of failure could be the unexpected presence of the repressor molecule, such as tetracycline, in the natural environment in a form and concentration that could successfully repress the lethal sterility genes in a fraction of escaped animals. Elevated concentrations of many biochemicals—antibiotics, caffeine, hormones, and pharmaceuticals—have been recorded in surface water bodies in the United States (Kolpin et al., 2002). Many manufactured biochemicals pass through domestic and industrial sewage and stormwater runoff systems in biologically active forms and then enter rivers, lakes, and coastal waters where they can remain in solution in the water column. It then must be determined whether those waters contain a bioactive compound that could repress sterility genes. Should that occur, biotechnologists would have to design a different genetic control system that responds to another compound not present in such quantities.

Default Gene Blocking by Interference RNA and Exogenous Rescue

It is possible to incorporate into transgenic constructs sequences that interfere with posttranscriptional expression of a target gene. This general strategy for blocking gene expression uses RNAi and could involve dsRNA (Fire et al., 1998), "hairpin" RNA, or other forms. RNAi would be used to block expression of an endogenous factor that is essential to development or reproduction. That compound would then be supplied exogenously in the diet to allow normal development or reproduction of captive GEOs. Upon escape from confinement the transgenic organisms would not survive or reproduce because necessary substance would no longer be available.

Recent advances in silencing diverse target genes in plants (Smith et al., 2000; Wesley et al., 2001; see Chapter 3) open the possibility of developing similar approaches in fish and shellfish: The transfer of intron-containing constructs encoding self-complementary "hairpin" RNA (ihpRNA) led to silencing of the targeted genes in 90–100% of individual plants and to a high degree of silencing within individuals; some plants exhibited almost complete knockout of the target gene. For utility in bioconfinement, it would be best to completely silence target gene expression in 100% of the transgenic individuals and to have confirmation of stable inheritance of the intact gene-silencing construct.

Strengths

Homologous recombination is not necessary for function, only knowledge of the sequence of the target genes in the target organisms.

Weaknesses

Experiments using RNAi in fish have not demonstrated sufficient specificity of target gene inhibition, and RNAi might not be suitable as a mechanism for knockdown or knockout of target gene expression . Not enough is known about the effects of this approach on whole fish and shellfish to develop an adequate list of strengths and weaknesses. Ten or more years of research could be needed to reach and verify effective bioconfinement.

Externally Administered Gene-Specific Compounds

It is possible that various nucleotide analogues could be used as gene-specific compounds to interrupt the expression of developmentally important genes. Those compounds could prevent development of reproductively necessary organs and gonad tissues, gametes, and other structures in the production organisms (not in the breeding stock) early in development but would not interfere with desirable characteristics of the target organisms. Candidate molecules include nonendogenous analogues of nucleotides that bind to a specific endogenous DNA sequence and interrupt its normal expression (Corey and Abrams, 2001; Ghosh and Iverson, 2000; Heasman, 2002). The targeted binding of those analogues relies on the normal specificity of base pairing, as occurs in the hybridization of naturally occurring nucleic acids. But the analogues have altered properties that are the result of chemical modifications of the backbone structure that supports the nucleotide bases. The altered backbone makes the analogues resistant to degradation in the target cell, more so than the RNA oligonucleotides involved in the gene-blocking approaches described above. The analogues have been used to shut down or knock down expression of specific genes (a compilation of the work using the most successful of these analogues can be found at the Gene Tools web site, http://www.gene-tools.com/). Thus the analogues could be adapted to bioconfinement by disrupting development of reproductively essential cells, tissues, or organs without altering desirable characteristics of the target organism.

Strengths and Weaknesses

This discussion could constitute the first proposal to apply gene analogues for bioconfinement, so consideration of strengths and weaknesses is highly speculative at this point. A description of the fundamental advantages of morpholino derivatives could give some indication of possible strengths of this general approach. A morpholino is an antisense oligonucleotide derived from the morpholine ring, which replaces the ribose or deoxyribose rings characteristic of RNA- and DNA-type oligonucleotides (http://www.gene-tools.com/Questions/body_questions.HTML). Gene Tools currently is the only licensed producer of these compounds in the United States. Mention in this publication does not confer endorsement of the firm or its products). Three general weaknesses would warrant attention if this approach were pursued for bioconfinement. First, the analogues could fail to perform as necessary. The analogues also could be too expensive for widespread use. Finally, the analogues could prove environmentally unstable and thereby present a hazard to nontransgenic organisms.

Gene Knockout

A target gene could be inactivated by knockout processes similar to those used to produce transgenic knockout mice. However, the process requires the ability to replace the target gene in the target organism with a knockout gene. This is additionally dependent on technology to allow the production and selection of chimeric progenitor animals through homologous recombination and selection of the desired genotypes.

Strengths and Weaknesses

This technology is now unavailable or insufficiently robust for application in a bioconfinement protocol (Rong and Golic, 2000, 2001; Rong et al., 2002).

Naturally Sterile Interspecific Hybrids

There are few well-documented examples of sterile hybrids among fish and shellfish. All the known examples involve hybrids between taxonomically distinct species, or interspecific hybrids (Chevassus, 1983).

One recent study reported highly effective achievement of "natural" sterility through two consecutive but different forms of interspecific hybridization (Liu et al., 2001). The first event, in the F_3–F_8 generations (female red crucian carp × male common carp), yielded tetraploids that apparently produce diploid (not haploid) gametes. The second hybridization mated a male F_3–F_8 hybrid (diploid sperm) with haploid eggs from a female of a third species, either Japanese crucian carp or Xingguo red carp. This yielded triploid fish, all of which were sterile. This is an ideal bioconfinement system: It provides 100% sterility of all progeny of the second cross, and it theoretically eliminates the need to screen for failed cases or to bear the added cost of artificially induced triploidy. The challenge is to find similar systems for "natural" bioconfinement across the spectrum of fish and shellfish species that have suitable characteristics for thriving in aquaculture systems, that consumers are willing to eat, and that aquaculturists are willing to produce.

Strengths

An interspecific hybrid clearly shown to be 100% sterile but viable and with suitable production characteristics would offer several bioconfinement strengths in production aquaculture: the highest possible reliability for a single confinement measure, ease of application, and obviation of the need for screening to remove potentially fertile individuals as required when relying on triploidization.

Weaknesses

Given that many interspecific hybrids of fish and shellfish are fertile, it is not safe to assume that any one hybrid is sterile without reliable evidence to the contrary (Thorgaard and Allen, 1992). Also, the degree of sterility in female and male hybrids might not be the same (Donaldson et al., 1993).

Combining Triploidization with Interspecific Hybrids

Where it is important to prevent reproduction by all individuals, Thorgaard and Allen (1992) proposed the use of interspecific triploid hybrids when the diploid hybrid is either unviable or fertile. Triploid hybrids involving some species of salmon and trout have higher survival rates than do their equivalent diploid hybrids (Chevassus, 1983), and they have been studied to a limited extent (Benfey, 1989). The triploid hybrid might be acceptable to aquaculture producers if it exhibits viability and good performance in other production traits, such as growth and general resistance to disease. Indeed, the combination of dramatically enhanced production traits in transgenic fish or shellfish with triploid interspecific hybridization to achieve confinement objectives might meet these conditions.

Strengths and Weaknesses

This approach involves strengths and weaknesses that are similar to those discussed above for sterilization via induction of triploidy. However, concerns about the adverse effects of escapees on wild relatives would apply to either or both parental species, depending on their co-occurrence in accessible ecosystems. Such concern would arise either because some undetected percentage of escapees is not functionally sterile or because sterile individuals enter into normal courtship behavior and can therefore disrupt the reproductive success of wild mates.

Abandoned and Inappropriate Methods

Efforts to render fish or shellfish sterile through surgery or chemical treatment have been abandoned for various reasons. Surgical removal of gonad tissue is the oldest method of sterilizing fish, starting with its use by the Chinese on farmed carp centuries ago and on salmon and trout species from the mid-1700s to the late 1900s (Donaldson et al., 1993). Under experienced hands, surgical sterilization can be effective, and it can offer high recovery rates. But it is not a serious candidate for commercial-scale bioconfinement of transgenic lines because of the cost of labor-intensive surgery and the need to wait until each fish has grown to at least 100g to exhibit gonads. Chemosterilization of fish, through treatment with mutagens, gonadotropin antagonists, antisteriod compounds, and androgens, as well as sterilization by X- or gamma-irradiation, has been abandoned because fish destined for human consumption would pose food safety concerns or be unacceptable to consumers (Donaldson et al., 1993). In mollusks and crustaceans, chemical shock to induce triploidy, and thus disrupt sexual reproduction, has shown mixed results (Dumas and Campos Ramos, 1999; Fast and Menavesta, 2000). Chemical shocking with cytochalasin B has been highly effective in some species, such as oysters, but it causes high mortality in others. The chemical 6-dimethyl-aminopurine (6-DMAP) appears to be less toxic for inducing triploidy, but it too produces triploidy in fewer than 100% of treated fish.

Interspecific hybridization which often fails to disrupt sexual reproduction is relatively common (Collares-Pereira, 1987; Turner, 1984), occurring, for instance in at least 56 fish families (Lagler et al., 1977). Because the majority of the known interspecific hybrids of fish and shellfish also are fertile, any new interspecific hybrid combinations that are tried as a bioconfinement measure should be thoroughly screened for evidence of fertility in both sexes.

Gene regulation strategies aimed at biological control of pest or nuisance species are not appropriate for bioconfinement. Consider for example the research under way in Australia to develop transgenic fish lines that bear a "daughterless gene" construct as a strategy for eradication of alien, nuisance fish species that have invaded river systems (CSIRO, 2002; Nowak, 2002; Woody, 2002). The strategy is inappropriate for bioconfinement of transgenic fish and shellfish in aquaculture because the aim is quite the opposite—it is to spread the daughterless gene construct as fully as possible into the alien, nuisance fish population. The general idea would be to release large numbers of alien species fish bearing the daughterless-gene construct among free-roaming individuals of the pest species and thus trigger a collapse of the pest population.

BIOCONFINEMENT OF INSECTS

As noted in the introduction to this chapter, there are many reasons for producing transgenic insects. It will be important, no matter the justification, to prevent those insects from going where they are not wanted and to prevent their transgenes from spreading to wild or domesticated populations.

Sterile Insect Technique

The sterile insect technique (SIT), originally developed for biological control of insect pests, also could be applied to biologically confine transgenic traits of insects. The traditional approach involves the release of mass-reared and sterilized male insects to mate with wild females, thus reducing the pest population (Braig and Yan, 2002; NRC, 2002b; Wimmer, 2003). Radiation is most commonly applied to colony-reared insects to create mutations that induce sterility. The amount of mutation is adjusted so that every gamete produced by mutagenized insects will contain at least one such lethal mutation. Chemical sterilants were evaluated to induce dominant, developmentally lethal mutations (Borkovec, 1975, 1976; Grover and Agarwal, 1980; Knipling, 1968). However, previously and currently available chemosterilants pose hazards to workers in mass-rearing factories, and available chemicals cannot be applied to the indigenous pest population without endangering nontarget species. Thus, ionizing radiation, most often from an isotopic source (^{60}Co, ^{137}Ce) or an electron accelerator tuned to produce hard X-rays, is used far more commonly.

In SIT for pest control, organisms are grown in a colony and then subjected to treatment that damages their gametes to the point at which no progeny of a mating with the treated insects can survive (Calvitti et al., 1997; Krafsur, 1998). The insects are then released to mate with their wild conspecifics. In the idealized case, all offspring of such mating will receive one copy of a dominant lethal gene. However, population control can be effected with less than absolute sterility. This has been successful in the codling moth *Carpocapsa pomonella,* where substerilizing doses of radiation created males with chromosomal translocations that reduced their fertility. *Lucilia cuprina* blowfly males were developed with translocations between the autosomes and the Y chromosome (Calvitti et al., 1998; Carpenter et al., 2001; de Azevedo et al., 1968; Gracia and Gonzalez, 1993; Hardee and Laster, 1996; Hasan, 1999; Kerremans and Franz, 1995; Makee and Saour, 1999; Mansour and Krafsur, 1991; McInnis et al., 1994; Qureshi et al., 1993; Seth and Sehgal, 1993). To use SIT for bioconfinement or confinement, sterility should be as close to 100% as possible. However, irradiation or other sterilants may damage the general vigor and competitiveness of the treated insects (Stiles et al., 1989). Thus, the use of SIT techniques as a confinement method may conflict with other intended uses, should exposure to sterilants result in a less competitive organism. This must be considered in evaluating SIT technology for bioconfinement of transgenic organisms. For example, should the effective sterilizing dose for a given insect cause a great deal of somatic damage, resulting in a less competitive insect, SIT would not be an effective method. In addition, use of sterilizing technology for bioconfinement would require rigorous quality assurance.

Means for ascertaining fertility of insects subsequent to exposure to sterilants of SIT insects do exist, although their successful implementation can depend heavily on species-specific behavior and biology (Katsoyannos et al., 1999; Lux and Gaggl, 1996). Additionally, many of the most effective methods require so much time or such destructive testing of target organisms that it would be unfeasible for a program involving large numbers of transgenic organisms to be biologically confined. It is more effective to establish doses of sterilants that cause the desired sterility and then to determine that target organisms receive this dose. Standard methods for the process are available (Committee, 2002).

Since its inception, SIT has been applied worldwide to a variety of insects (Table 4-2; Van der Vloedt and Klassen, 1991) indicating that reorientation of this approach to achieve bioconfinement of transgenic insects (rather than to control pest insects) would be possible for a broad range of species.

TABLE 4-2 Insects Subjected to the Sterile Insect Technique

Insect	1991 sites	Previous sites
Screwworm	Guatemala, Belize, Libya	Curaçao, U.S., Mexico, Puerto Rico, U.S. Virgin Islands
Mediterranean fruit fly	Guatemala, U.S. (Hawaii)	Italy, Peru, Mexico, U.S. (California), Israel
Caribbean fruit fly	U.S. (Florida) fly-free zone	U.S. (Florida)
Melon fly		Japan
Oriental fruit fly	Japan, Brazil	Mariana Islands (Rota), U.S. (Hawaii)
Onion fly	Netherlands	Netherlands control
Mexican fruit fly	U.S., Mexico	U.S., Mexico (quarantine + fly-free zone)
Cherry fruit fly	Switzerland	
Tsetse fly (4 species)	United Republic of Tanzania, Nigeria, Nigeria, Zanzibar, Burkina Faso	
Sheep blowfly	Australia	
Tobacco budworm		U.S.
Stable fly	U.S. Virgin Islands (St. Croix)	
Tsetse fly	United Republic of Tanzania, Nigeria, Zanzibar	

SOURCE: Adapted from Van der Vloedt and Klassan, 1991.

The first field release that combined SIT and genetic engineering applied the latter principally to monitor the effectiveness of SIT, but had the corollary effect of confining the transgenes. It involved release of sterile transgenic pink bollworms—a lepidopteran pest of cotton—that bore the marker gene for green fluorescent protein (GFP) as part of a biological control program run by the U.S. Department of Agriculture in the cotton-growing areas of Arizona (Staten et al., 2001). Under ultraviolet light, GFP, even in dead insects, allows visual discrimination of sterile from fertile native bollworms (Braig and Yan, 2002). It should be possible to apply traditional SIT principally to prevent movement of transgenes into wild insect populations, rather than as a biocontrol method for a pest insect.

Strengths

The techniques developed for pest control that rely on induction of sterility or partial sterility can prevent flow of genetic material into conspecific populations (Marsula and Wissel, 1994; Robinson, 2002). SIT produces infertility through induction of mutation. Ideally, the treatment does not interfere with the desired characteristics of the target organism.

Weaknesses

Failure of SIT for bioconfinement of transgenic insects in large-scale applications would result from inadequate sterilization in the mass-reared insect population and subsequent release of fertile insects. The rates of sterility, in terms of fertile offspring of steriles in practice, vary from effectively 100% to 75%, depending on the target organism. Sterility in Dipteran flies is usually high, effectively 100%, whereas in other insects the sterility can be lower and still be effective in pest control. Thus the use of SIT in bioconfinement must consider the response of the target organism to the sterilizing method.

Transgenic Sterile Insects

Gene transfer also has been proposed as a way to produce sterile insects for biological control that would improve on the traditional SIT approach and replace the use of radiation or other mutagens to induce sterility (Alphey, 2002; Alphey and Andreasen, 2002; Thomas et al., 2000; Wimmer, 2003). An important motivator for this line of research is that radiation-based SIT tends to depress the vigor and competitive ability of sterile males, thus undermining SIT's effectiveness for biological control. This also will be a concern if SIT approaches are applied to biologically confine transgenic insects. Ideally, bioconfinement methods would abrogate reproduction without altering any other desirable traits.

One strategy involves developing transgenic traits for inducible genetic sterility. An example demonstrated in fruit flies involves a transgene-based dominant embryonic lethality system that can generate large quantities of competitive but sterile insects (Horn and Wimmer, 2003). The sterile insects are vigorous adults but their transgenes cause lethality after transmission to progeny. This embryonic lethality can be suppressed maternally in the laboratory in order to propagate the strains.

Transgene-based embryonic lethality can combine with another strategy involving transgenic female-specific lethality systems to produce sterile males (Heinrich and Scott, 2000; Thomas et al., 2000). Female-specific lethality can be turned on and off through inclusion of a tetracycline-activated regulatory element in the transgenic construct. The construct can be suppressed by supplementing food with tetracycline during insect rearing in captivity.

Strengths and Weaknesses

Almost nothing is known about the strengths or weaknesses of these transgenic methods for bioconfinement because scientists are at least 10 years away from application. As has been the case for traditional SIT, the rates of sterility, in terms of fertile offspring of steriles, will likely vary from effectively 100% to 75% depending on the target organism and the specific transgenic method used.

Ecological Characteristics of Production Site

For commercially important and partly or wholly domesticated transgenic insects, the amount of confinement needed depends strongly on the insect's biology. In the case of silkworms, little or no confinement should be necessary because the insects are completely adapted to commercial silk production, so they cannot escape. However, low vagility (mobility) cannot be expected should transgenic honeybees be produced, because of the possibility that transgenic bees would mate with wild-type bees of the same species.

Climatic or ecological conditions in some places should provide confinement for transgenic insects, depending on the insect's ecology and behavior and on the feasibility of keeping it confined to that region. However, inadvertent or purposeful transport to a more suitable area could easily abrogate such confinement. For example, the Mediterranean fruit fly (medfly) and other tropical insects would have no chance of survival in the immediately accessible environment should they escape from a rearing facility located in an area with a cold climate or lacking appropriate hosts. Although the diverse diet of the medfly makes this latter confinement approach problematic, it could be implemented where the insect in question has a highly restricted host range.

Fitness Reduction and Regulation of Gene Expression

Some transgenic, mass-reared insects that serve as biological factories to produce valuable proteins could escape confinement and interbreed with their wild specifics. For instance, medflies and pink bollworms can be engineered to produce valuable transgenic proteins (Peloquin and Miller, 2000). Although it is unlikely that a medically or industrially important protein produced by such a transgenic insect would confer any selective advantage, it is not, at the very least, good environmental hygiene to allow the escape of such a transgenic insect. Perhaps such biological factory insects could be rendered flightless or incapable of long-range dispersal by use of a flight-defective mutation, such as the long-known recessive *Drosophila* gene *vg,* which results in flightless insects. Alternatively, technology for gene blocking or gene knockout in development for bioconfinement of transgenic fish and shellfish might be developed to prevent reproduction or postescape survival of industrial transgenic insects.

5

Bioconfinement of
Viruses, Bacteria, and Other Microbes

INTRODUCTION

The use of genetically engineered microbes can offer enormous potential benefits. Because viruses, bacteria, and fungi are natural pathogens of insects and other pests, microbes can be harnessed and genetically enhanced as agents of biocontrol. Bacteria and fungi also are able to degrade some environmental pollutants, and molecular technology allows us to expand the list to include other toxic compounds as well. There is a long human history of using microbes in agriculture, food processing, waste treatment, and other beneficial capacities. Modern molecular methods allow us to broaden the range of useful applications, and all of the evidence indicates that the methods used to generate genetically engineered organisms (GEOs) are not intrinsically dangerous. Some caution is warranted, however, because information about the ecology and evolution of transgenic microbes in the wild is limited. Microbes occur in extremely large populations with short generation times, so they adapt quickly to adverse conditions. Their environments change constantly, resulting in unpredictable and variable selection pressures. Bacteria also can transfer DNA into unrelated microbes, and the long-term ecological consequences of that transfer are unclear (Bushman, 2002). The consequences of releasing transgenic microbes into the environment have not been evaluated adequately. As with the plants and animals discussed in the earlier chapters, it is impossible to generally predict the fitness consequences of genetically engineering a microbe. The new genetic combination could be beneficial or deleterious for a microbe's survival in the wild, depending on the ecological context (e.g., through interactions with the environment and with other microbes). Most studies show that genetically engineered microbes are relatively less fit than are their nonengineered counterparts, although that can be a faulty assumption. Each case of genetic engineering must be considered on its own.

This chapter is on the bioconfinement of genetically engineered microbes, especially bacteria, fungi, and viruses. The potential effects and need for bioconfinement in microbes are discussed first. Then there is a section that identifies and describes the major methods of bioconfinement for bacteria, fungi, and viruses and discusses the strengths and weaknesses of each method. Next are considered the effectiveness of those methods in different spatial and temporal scales and their potential effects on biological populations and ecosystems. The needs, feasibilities, and realities of monitoring, detecting and culling genetically engineered bacteria, fungi, and viruses are discussed. Finally, the aforementioned topics are related to the bioconfinement of microalgae.

Because the committee was not specifically asked to evaluate bioconfinement techniques for microbes, this chapter is purposefully less substantive than are the chapters on plants and animals. However, the widespread appreciation for the usefulness of transgenic microbes warrants their treatment here. Earlier NAS reports also have dealt with genetically engineered microbes (NRC, 1989a, 2002b).

The 1989 NAS report, *Field Testing Genetically Modified Organisms,* extensively evaluated environmental risks associated with the release of transgenic microorganisms. The recommendations in that publication influence policy decisions today, despite the advancement of molecular technology in the intervening years. Although portions of the present report discuss transgenic microbes, the subject cannot be dealt with in detail because microbes were not a central focus of the committee's charge. The committee suggests that genetically engineered microbes be reconsidered on their own.

POTENTIAL EFFECTS OR CONCERNS, AND NEED FOR BIOCONFINEMENT IN VIRUSES, FUNGI, AND BACTERIA

The three potential areas of concern that attend the release of genetically engineered bacteria, fungi, and viruses are similar to those for any other class of GEO: invasion, displacement, and transfer. Together they can be used to argue that bioconfinement measures should be considered.

Invasion into Indigenous Populations

Viruses

Genetically engineered viruses could infect and harm nontarget hosts. Because viruses are obligate intracellular parasites, they require metabolizing (living) cells to replicate their genomes and make progeny. The reliance on host cells often produces strong selection for viruses to evolve more efficient mechanisms to exploit their hosts. In turn, selection of the host favors genotypes that excel in their ability to repel virus attack. Viruses can gain the upper hand in these coevolutionary battles simply because they can evolve more rapidly than do their hosts (Levin and Lenski, 1983). Thus, a virus can be successful by evolving a greater propensity to exploit the host, through creating more progeny per infection per unit time. Alternatively, viruses could be evolutionarily successful by adapting to infect a greater variety of hosts (DeFilippis and Villarreal, 2000). The latter adaptation exemplifies the potential consequence of releasing genetically engineered viruses into an ecological community of naïve (inexperienced) hosts. The concern is not the introduction of engineered alleles (such as transgenes) per se, but that the foreign strain of virus will harm a nontarget host species that is ill-prepared to defend against the viral attack because of its lack of resistance genes. Support for this idea comes from studies that demonstrate elevated virulence in naïve host populations (Bull, 1994; Taylor et al., 2001). Thus, releasing genetically engineered viruses into the environment might result in their becoming successfully established in nontarget hosts.

Fungi

The host range of fungi also can evolve. Reports of changes in the known host range of plant pathogenic fungi are common (Mundt, 1995). For example, the scabrum rust, which is pathogenic on *Agropyron scabrum* and *Hordeum vulgare* (barley) in Australia, arose from a cross between *Puccinia graminis* f. sp. tritici (wheat stem rust) and *P. graminis* f. sp. secalis (rye stem rust) (Burdon et al., 1981). In 1991 cultivated barley was found to be heavily infected with a new variety of *P. coronata*, a fungus that was known for more than 200 years to cause serious disease in cultivated oat.

From these and other examples, it appears that the pathogen can genetically alter its host range. However, studies on the rice blast fungus showed that many genes are required for the fungus to attain the high fitness required for field survival on a new host (Valent et al., 1991). Those results indicate that survival in the wild in the face of competition from other organisms and changing environmental conditions can be far more demanding than surviving in a laboratory under optimal conditions. Field studies have been conducted on a transgenic mycoinsecticide to monitor the fate of the fungi under field conditions (Hu and St. Leger, 2002). The fungus was released onto a plot of cabbage and survivorship was determined in nonrhizosphere and rhizosphere soils. In nonrhizosphere soils, the fungal propagules decreased from 10^5 to 10^3 per gram after several months. However, recombinant fungi engineered only with the gene for green fluorescent protein (GFP) remained at 10^5 propagules in the rhizosphere soil. Fungi that contained an additional protease gene did not persist as well in the rhizosphere as did the GFP genotypes. The observations are consistent with what has been observed for transgenic bacteria—that adding genes to cells often decreases the microbes' fitness, but that some ecological contexts (such as placement in the rhizosphere) can promote their survival.

Bacteria

In theory, genetically engineered bacteria introduced into the environment can become established in a microbial community. A limited number of studies to assess this possibility have been done on bacteria in aquatic and terrestrial environments. In one study (Scanferlato et al., 1989) the viability of genetically engineered and wild-type strains of *Erwinia carotovora* were compared after their addition to an aquatic microcosm. Both declined in viability at the same rate, and within 32 days neither was detectable by viable counts. Those data suggest that the newly introduced bacteria, whether genetically engineered or wild-type, were poorly adapted to the new environment and therefore were unable to compete with indigenous species. The observations are consistent with theory and with laboratory experiments on resource competition. Both suggest that the competitor that grows at the lowest concentration of a limiting resource will survive and thereby displace all inferior competitors (Hansen and Hubbell, 1980; Tilman, 1982).

In nature, most nutrients are present at low concentrations in terrestrial and aquatic environments (Madigan et al., 2003); therefore, only those microorganisms that can compete for those limited resources would be expected to thrive. To overcome the likelihood of introduced microbes, being poor competitors, the genetic capabilities to perform a particular function are commonly introduced into bacteria that already are adapted to a particular habitat (Glandorf et al., 2001). For example, bacteria are being used for bioremediation of oil-contaminated beaches and polluted soils. The common practice is to apply nutrients, such as nitrate and phosphate, to the contaminated area to promote growth of indigenous bacterial populations, which likely will include microbes that can metabolize the pollutants.

As a rule, introducing genes into indigenous bacteria to perform a specific function is preferable to introducing exotic bacteria that contain those genes. Although bacteria initially could be maladapted to a new environment, they often can change and increase their growth rate significantly. In one study, transgenic bacteria were incubated in lake water for 15 days and then reisolated. The growth rates of the reisolates were more than 50% higher in the lake water when compared to the original strain (Sobecky et al., 1992). The researchers concluded that bacteria can adapt to oligotrophic environments and the fitness of GEOs for survival can increase in aquatic ecosystems. The ability of bacteria to adapt to new environments over time is an important concern in their release to the environment. That adaptability also applies to non-GEOs, though introduction of non-transgenic exotic bacteria into new environments has not raised strong concerns.

In another study, the viability of genetically engineered *Pseudomonas putida* that contained a gene for the synthesis of the fungal inhibitor phenazine was compared with its wild-type parent in the rhizosphere of wheat plants for two growing seasons (Bakker et al., 2002). In both seasons, the genetically engineered and the wild-type strains decreased to below detectability within a month after the wheat harvest, indicating that the rhizosphere was essential to the survival of the introduced bacteria. In one season, within days of sowing the genetically engineered strain decreased more rapidly than did the wild-type strain. In another growing season, however, no difference in density was observed for the two strains, indicating that the additional metabolic load on the GEO did not reduce its ecological fitness. Those results over successive years suggest that ecological fitness, at least in soil, depends on the variable environmental conditions encountered by the GEOs.

Considerable data suggest that the increased genetic load that results from introducing additional genes into a microbe usually reduces its growth rate unless a strong selection pressure favors the added genes (Lenski and Nguyen, 1988; Milks et al., 2001; Zund and Lebek, 1980). The observed decrease in growth rate apparently can result from the additional products synthesized from the DNA rather than from the replication of the DNA (Lenski and Nguyen, 1988). However, several research groups have reported that a genetically engineered bacterium can grow at the same rate as or even faster than its parent does (Bouma and Lenski, 1988; Devanas and Stotzky, 1986; Edlin et al., 1984; Hartl et al., 1983; Marshall et al., 1988). Those latter observations are not well understood, and it is unclear whether the organisms as grown in the laboratory would be ecologically fit in a natural environment. Also, cases have been reported in which genetically engineered bacteria coexist with indigenous populations (Kargatova et al., 2001). If the introduced bacteria were resistant to an antibiotic present in the environment, in theory it should thrive because of reduced competition from susceptible strains. Thus, the assumption that all genetically engineered bacteria are unfit in natural environments cannot be sustained. However, the same would be true for exotic non-GEOs.

Displacement of Indigenous Populations

Viruses

It is theoretically possible that genetically engineered viruses could displace resident species. In theory, the coevolutionary battle between viruses and their hosts leads to a never-ending arms race; hosts evolve resistance, viruses evolve counterresistance, and the cycle repeats with both species constantly running to remain in place (this is the "Red Queen hypothesis"; Clarke et al., 1994). Whereas the hosts and viruses must continuously evolve to maintain same dual-species interaction, at least this scenario produces long-term stability of biodiversity in the ecosystem. In contrast, a newly introduced virus that can prove more virulent in nontarget host tips the balance in its favor (Bull, 1994; Taylor et al., 2001). Subsequent devastation of the nontarget host is the primary concern, but a separate concern is that the resident virus could lose out because it is less efficient (relatively less virulent). Extinction of the endemic virus could disrupt the ecological community; for instance, introduction of the relatively more virulent species could force the nontarget host to a lower equilibrium density in the community, producing a cascade effect elsewhere in a food web. Viruses might be underappreciated in terms of their influence on regulating large-scale ecosystem processes (Fuhrman, 1999).

Bacteria

Several studies have reported on the effect of adding genetically engineered or wild-type bacteria to resident flora in aquatic and terrestrial environments. In one study (Scanferlato et al., 1989), genetically engineered strains of *E. carotovora* were added to an aquatic microcosm, and the effects were measured in some elements of the indigenous population. Thirty-two days after inoculation the number of total and proteolytic bacteria was the same in the inoculated and uninoculated microcosms. Neither did the inoculation affect the number of amylolytic and pectolytic bacteria in the water or sediment.

In another study, genetically engineered *Pseudomonas fluorescens* were released into indigenous populations near wheat plants (De Leij et al., 1995). The results for culturable organisms can be summarized as follows: *P. fluorescens* and the unmodified strains produced the same results, the perturbations to the microbial population were small, and the release of bacteria had no obvious effect on either plant growth or health.

Genetically engineered *P. putida* that contained a gene for the synthesis of the fungal inhibitor phenazine and its wild-type parent were added to the rhizosphere of wheat plants (Bakker et al., 2002). Neither the transgenic strain nor its parent affected the metabolic activity of the soil microbial population, and only transient changes were observed in the composition of the rhizosphere fungal microflora. Although the GEO had the greater effect, the authors suggested that the effect of the introduced GEOs was only minor in comparison to those that result from such common agricultural practices as plowing or crop rotation.

Most studies that have assessed the influence of genetically engineered microbes on microbial populations have studied effects on culturable organisms alone. Yet less than 1% can be grown in culture (Madigan et al., 2003). However, polymerase chain reaction technology and measurement of ribosomal DNA (rDNA) patterns make it possible to analyze entire bacterial populations. Using those techniques, Robleto and colleagues (1998) showed that introduction of engineered strains of *Rhizobium* synthesizing a narrow-spectrum-peptide antibiotic reduced the diversity of α-Proteobacteria; while the total bacterial population was not substantially affected. In another study *P. putida,* genetically engineered for increased activity against soilborne bacterial and fungal pathogens, were released into the rhizosphere of wheat and their effect on indigenous microflora was determined (Bakker et al., 2002). Effects of the genetically engineered bacteria on the rhizosphere fungi and bacteria were analyzed, using amplified ribosomal DNA restriction analysis. A transient change in the composition of the rhizosphere was noted, but several soil microbial activities, such as soil nitrification and cellulose decomposition, were unaffected. The limited data from all of these experiments indicate that the introduction of genetically engineered microorganisms has mostly transitory effects on indigenous populations that are unlikely to be significant in the field. The effects of adding transgenic microbes are not likely to be any greater than are those that attend the addition of nontransgenic species.

To reduce the possibility of changes in the indigenous microbial population that result from the release of genetically engineered microbes, the committee advises that strains be used that are likely to be poor competitors in the local environment. The limited available data suggest that the introduction of genetically engineered microbes into the environment is unlikely to have significant long-lasting effects on microbial communities.

Horizontal Genetic Transfer into Local Populations

A third concern of introducing genetically engineered microbes is the potential consequence of horizontal transfer of engineered genes from introduced microbes into local populations.

Viruses

Viruses have two distinct mechanisms for the exchange of genetic material. When two or more DNA or RNA viruses infect the same host cell, recombination can lead to hybrid progeny that contain genetic information from both parents (Hershey and Rotman, 1949). In contrast, some RNA viruses have genomes that are split into several smaller segments, and co-infection can produce hybrids that contain a random reassortment of the segments found in the infecting parent viruses. Such exchanges have periodically led to new strains of influenza virus that have caused human pandemics (Palese, 1984; Webby and Webster, 2003).

Recombination in viruses, can promote linkage equilibrium (free association of alleles) to create the potential for engineered alleles to enter and circulate within a local gene pool. Laboratory experiments show that gene exchange can profoundly affect virus evolution (Rambaut et al., 2004; Turner, 2003; Turner and Chao, 1998), and it is generally accepted that viral recombination in natural infections is a major force in the evolution of new viruses (e.g., Goldbach, 1986). Many viruses also can recombine with host chromosomes, thus introducing virus-derived genes into the host genome and the host gene pool. The fitness effects of engineered genes in the original virus background are likely to be assessed before strains are released into target populations of hosts. The concern is that those genes could have unanticipated effects when they transfer horizontally from the engineered background into new ones. Epistasis (gene interaction) between introduced alleles and those in the gene pools of other species can hamper the fitness of individuals in those groups.

Hammond and colleagues (1999) studied transfer of engineered genes through virus recombination and considered the likelihood that the process would harm natural populations. The greatest concern identified by those authors was the creation of new virus types as a result of recombination between wild-type viruses and unrelated transgenes in genetically engineered plants. However, the rate at which this happens is unlikely to exceed that in naturally occurring mixed infections of viruses of nonengineered plants. More data are needed from field trials to evaluate the benefits and risks associated with release of transgenic viruses. The most extensive survey to date (Thomas et al., 1998) studied interactions between transgenes derived from potato leafroll virus and viruses to which transgenic plants were exposed. The experiments revealed no evidence of recombination, altered transmission, or altered virus properties, suggesting that such phenomena are extremely rare.

Bacteria

Most species of bacteria have several mechanisms for horizontal gene transfer: DNA-mediated transformation, in which "naked" DNA is transferred to recipient cells; generalized or specialized transduction, in which donor DNA is enclosed in the coat of a bacteriophage; and conjugation, in which DNA—primarily through plasmids—is transferred from donor to recipient cells after contact between the two. Bacterial gene exchange can affect the persistence of a strain or its engineered alleles, and a genetically engineered bacterium can be the donor or the recipient of genetic information by horizontal gene transfer. Only a new genetic combination with higher fitness than the indigenous genotype under the specific environmental conditions has a high likelihood of persisting (Koonin et al., 2001).

In the laboratory, the transfer of genetic material is most efficient between members of the same species. However, transfer in nature has been observed between widely different genera, and even domains, of bacteria. Comparative analysis of bacterial, archeal, and eukaryotic genomes suggests that a significant fraction of the genes in prokaryotic genomes have been involved in horizontal transfer over evolutionary time (Koonin et al., 2001). At least one bacterium, *Agrobacterium*, can transfer DNA into plants, and it is the workhorse of plant genetic engineering (Chilton et al., 1977). The broad-host-range plasmid RSF1010, when in *Agrobacterium*, can mediate its own transfer into plants as well as into other Gram-negative bacteria (Buchanan-Wollaston et al., 1987). Thus, plants have ready access to the gene pool of Gram-negative bacteria, thereby expanding the possibilities of horizontal gene transfer from prokaryotes to eukaryotes. Many varieties of tobacco (*Nicotiana tabacum*) contain genes transferred from *Agrobacterium* over evolutionary time (Furner et al., 1986). In the laboratory, *Agrobacterium* can transfer DNA into a variety of fungi (Bundock et al., 1995; de Groot et al., 1998; Piers et al., 1996) and because *Agrobacterium* and many fungi occupy the same habitat in soil, transfer between the bacterium and fungi might also occur in nature. *Escherichia coli* (*E. coli*) can transfer plasmid DNA into yeast in the laboratory (Heinemann and Sprague, 1989). It also has been reported that *Agrobacterium* can transfer DNA into mammalian cells (Kunik et al., 2001) and that *E. coli* can transfer plasmids into mammalian cells (Waters, 2001).

Several studies have examined the possibility of gene transfer from transgenic plants to bacteria (Gebhard and Smalla, 1999; Schlüter et al., 1995). The conclusion has been that such an occurrence would be extremely rare, although plant DNA can persist in the soil under field conditions for up to 2 years. Nielsen and colleagues (1998) emphasized that, although gene transfer can be a rare event, it is critical to understand the selective forces that act on the outcome of any transfer.

Plasmids are a common vector for cloning and moving transgenes from one organism to another, and conjugation in particular would allow these genes to move easily into other bacteria in the environment. Some conjugative plasmids are highly promiscuous in their ability to transfer horizontally between unrelated bacteria, and that has contributed to widespread dissemination of antibiotic-resistance genes, which often reside on plasmids. In contrast, other plasmids are nonconjugative and thus unable to be transferred. However, they can be transferred if another plasmid in the same cell provides the missing functions. The probability of horizontal transfer can be reduced by cloning genes into nonconjugative plasmids and by using bacteria that contain no other plasmids. The risk of transfer can be further minimized if the engineered genes are integrated into the bacterial chromosome rather than remaining on a plasmid. Further, the enzyme transposase—required for gene movement inside a cell, and frequently used in constructing transgenic bacteria—should be disabled. All introduced plasmids should be defective in conjugation functions. If any antibiotic resistance loci are used to mark strains, the resistance loci should not involve antibiotics that currently are in clinical use.

A sensible choice should be made for the bacterial strain introduced into the environment to carry out a specific function. For example, it would be unwise to introduce a close relative of a disease-causing bacterium because nonpathogenic relatives conceivably could differ only by the presence of a genetic element (plasmid, transposon, prophage) on which virulence genes reside. Thus, in theory, inadvertent acquisition of one or more functions might convert the introduced strain into a dangerous pathogen of humans, animals, or plants. This is because bacteria, unlike viruses, are not obligate parasites. That is, bacterial pathogenicity depends on an array of characteristics that relatively few bacteria have acquired through extended coevolution with a particular host (Salyers and Whitt, 2002). Therefore, it is unlikely that minor genetic modifications would convert a nonpathogenic strain into a pathogen, and laboratory experiments to achieve it—at least with *E. coli*—have thus far been unsuccessful (S. Moseley, University of Washington, personal communication, 2003).

Fungi

Evidence is weaker for horizontal gene transfer between fungi than it is between bacteria (Rosewich and Kistler, 2000). In one case, however, sequence analysis of a gene for chymotrypsin synthesis in a fungus in which chymotrypsins had never been observed revealed that the sequence is related to that of a soil bacterium (Screen and St. Leger, 2000). As more sequences become available, it should become clearer whether horizontal gene transfer can occur.

To reduce the possibility of horizontal transfer of genetically engineered alleles, it is advisable to use microbial strains in which transgenes are integrated into chromosomes. The enzyme transposase, required for gene movement inside a cell and frequently used in constructing transgenic bacteria, should be disabled. All introduced plasmids should be defective in conjugation functions. If any antibiotic resistance loci are used to mark strains, the resistance loci should not involve antibiotics that are currently in clinical use.

BIOCONFINEMENT OF BACTERIA, VIRUSES, AND FUNGI

Because they are small, easily dispersed, and numerous, genetically engineered microbes will require bioconfinement approaches that are different from those for other GEOs. Control centers on fitness reduction.

Fitness Reduction

Phenotypic Handicapping

One potential consequence of releasing transgenic microbes to the environment is that they could perpetuate by invading or displacing natural populations in competition for resources. The limited experimental data suggest that, in general, genetically engineered bacteria and viruses will be competitively less fit than their wild-type counterparts because of burdens associated with carrying and expressing additional functions coded by transgenes. As noted already, microbes generally fare poorly when introduced into a new environment, although numerous cases could be cited in which the genetically engineered microbe did not appear to be significantly handicapped compared with the parental strain (Bouma and Lenski, 1988; Devanas and Stotzky, 1986; Hartl et al., 1983; Marshall et al., 1988). One solution to the uncertainty of whether transgenic microbes or their genes will persist is to use strains with phenotypic handicaps, such as reduced survival capability, reduced reproductive capacity, low resistance to a predictable change in the environment (such as seasonal heat or cold), or a tendency to lose the specific function of concern (NRC, 1989a).

Viruses

Phenotypic handicapping is widely applied in the design of live viruses for use as vaccines (Murphy and Chanock, 2001). They do not cause disease but they stimulate the host's immune system to produce antibodies against wild-type viruses to fight subsequent infection. An ideal live vaccine is an attenuated (weakened) form of the virus that is phenotypically handicapped, thus ensuring that its competitive inferiority will prevent it from persisting in nature. That is, phenotypic handicapping hampers the vaccine or its engineered alleles from influencing the evolution of wild populations through entry of those elements into the natural gene pool. Similarly, attenuated viruses can be engineered to carry proteins of unrelated virus pathogens, and those chimeric vectors elicit immune responses without posing the threat of long-term persistence (Rose et al., 2001).

Recent studies have attempted to engineer herpes simplex virus-1 and other well-described viruses for therapeutic interventions, such as combating cancer through gene therapy (e.g., Advani et al., 2002). Overall, few data exist regarding the long-term persistence of genetically engineered viruses. The effectiveness of phenotypic handicapping of viruses as a confinement measure is not clear.

Bacteria

Phenotypically handicapped live bacteria also have been used in inducing cellular immune response (Stocker, 1990). When a gene that codes for a particular epitope (a short, linear peptide sequence that is a portion of a larger protein antigen) was inserted into a gene that codes for flagellin in *Salmonella* auxotrophic for aromatic acids, a cellular immune response to the epitope was generated (Verma et al., 1995). The bacteria did not multiply because of a lack of required compounds in the environment. Phenotypic handicapping of bacteria as a confinement measure already has been alluded to: One form involves the rapid decline of nonindigenous microbial strains (including genetically altered ones) after they are introduced into soil or aquatic environments (e.g., Glandorf et al., 2001; Scanferlato et al., 1989). The data support the widely accepted view that long-established microbial communities are able to resist invasion by foreign organisms (Liang et al., 1982).

However, one challenge to phenotypic handicapping is the evidence that genetically engineered strains can persist for long periods by quickly adapting to a local environment. For instance, Kargatova and colleagues (2001) observed that recombinant *E. coli* strains can persist for one year or more in aquatic microcosms, and that they can coexist with indigenous microflora. The strains adapted by decreasing their expression of cloned genes, suggesting that the genetically engineered bacteria tended to lose the genes of concern. Similarly, addition of plasmids does not necessarily lead to a long-lived handicap to the bacterial host. Bacteria can adapt through the mutation of genes that are not associated with the plasmid and thereby restore their growth rate to that of the original parental strain (e.g., Bouma and Lenski, 1988; Hartl et al., 1983).

Most experiments on phenotypic handicapping have been performed in the laboratory. But in natural environments such as lake water, the fitness consequences of added genetic material are more difficult to evaluate and apparently depend on many factors associated with the genetics of the bacteria and the environment, with its usually obscure and variable selection pressures. For instance, in a study involving prototrophic strains of *P. putida*, one with and the other without a plasmid, the plasmid-bearing strain was maintained in a lake system over a period of 2 months (Sobecky et al., 1992). The plasmid was lost within 24 hours if the strains were amino acid auxotrophs. Variations in weather, such as rainfall and temperature, can affect the population density of transgenic microbes introduced into the soil (Glandorff et al., 2001). In addition, it is possible that an altered microbe can become immediately more fit than the wild-type if the phenotypic change increases resistance to a noxious substance in the environment or increases the ability of the microorganism to metabolize a substrate in the environment. Expression of additional functions and their effects on fitness reduction have not been clearly defined, and a greater effort to examine this phenomenon in field tests is warranted (e.g., Palmer et al., 1997).

A second identifiable complication of phenotypic handicapping is that indigenous bacteria generally do not exist as free-living, individual cells. Rather, in their natural environment, bacteria often form highly structured clumps, called biofilms, with properties that are quite different from those of bacteria growing in the laboratory. For this reason, phenotypic handicapping of transgenic bacteria growing in liquid medium in the laboratory might not be relevant to performance in a natural setting. In particular, bacteria in biofilms are far more resistant to noxious chemicals, including antibiotics and heavy metals, than are individual cells (Madigan et al., 2003). Biofilms attach to inanimate objects in the environment and their formation requires the action of several genes. It could be undesirable for genetically engineered bacteria to persist long term, but they must live long enough to perform the intended function. Unless those bacteria form biofilms by attaching to such objects as rocks, soil particles, and teeth, they could be washed away by rain or other fluids. Several genes have been identified in *E. coli* and other microorganisms that are necessary for biofilm formation, so mutations in those genes could debilitate the organisms to the extent that they would not persist even for short periods with the indigenous flora (O'Toole et al., 2000; Pratt and Kolter, 1998).

A third difficulty of phenotypic handicapping is that the GEO could be so handicapped that it is not practical to use. Perhaps the best example is strain $\chi 1776$ (see Box 5-1).

BOX 5-1
χ1776

The most extreme case of phenotypically handicapping a microbe was carried out in the laboratory of Roy Curtiss after the 1975 International Conference on Recombinant DNA Molecules. The idea was to disable the K-12 strain of *E. coli*, considered safest for use in cloning experiments, and make it even safer, against the possibility of its escape from the laboratory. The thinking that went into the disabling process serves as a guideline for genetically handicapping other organisms.

Curtiss and his colleagues (1977) introduced mutations that precluded colonization of and survival in the intestinal tract, prevented biosynthesis of the rigid layer of the bacterial cell wall in nonlaboratory conditions, led to the degradation of DNA in any organisms that escaped the laboratory environment, permitted monitoring of strains, and inactivated DNA repair mechanisms. The key mutations were deletions or independent mutations. Thus, the traits were stable and unlikely to revert. The strain had a generation time twice to four times longer than the wild-type *E. coli* K-12 strain and likely would not compete well with healthy microbes in the environment. The strain also was resistant to most known *E. coli*-transducing phages and defective in inheriting many conjugative plasmids. Thus, the strain could not transmit genetic information by transduction or conjugation at detectable frequencies.

To celebrate the nation's bicentennial, the strain was named χ1776. It was used for the industrial production of insulin after the cloning of the eukaryotic insulin gene into the nonconjugative plasmid pSC101. However, the strain, with its multiple auxotrophic markers, sensitivity to detergents, and increased generation time, proved so difficult to grow that widespread use was clearly impractical. Accordingly, as studies with recombinant DNA became more routine and the guidelines for biocontainment were relaxed, wild-type strains of *E. coli* K-12 or HB101 were used as the hosts for DNA cloning. χ1776 is now just a memory of a bygone era.

Fungi

Two methods have been proposed to phenotypically handicap fungi. One is to isolate auxotrophic mutants that can exist on the pest host, in the case of a biocontrol agent, but that would not survive outside the host. Such mutants should be isolated using a physical mutagen, such as gamma or neutron radiation, that fragments genes and thereby prevents reversion. Another proposed technique is to render the fungi asporogenic (unable to produce spores), thereby helping not only to prevent their spread but also inhibiting the formation of dormant resting structures that resist heat, cold, desiccation, and other harsh environmental conditions (Gressel, 2001). Asporogenic mutants would be handicapped both in persistence and in the major structures dispersed by wind, water, or animals. Spores of some fungi, however, are required for pathogenesis. Thus, if the fungus is to be used in biocontrol, an asporogenic mutant would not be suitable, and this form of handicapping would be inappropriate.

Suicide Genes

Suicide genes can be used to confine bacteria and fungi under two circumstances. The first ensures that bacteria growing in a closed container (such as a vat) are unable to survive if they escape. The strain should die as quickly as possible after escape. In many situations this is best achieved by chemical sterilants. The second circumstance is to combat a perceived threat to the environment should released microorganisms persist beyond the intended period of usefulness, for example, for bioremediation of Superfund sites or for biocontrol of plant pests in agriculture. In that case, the GEO must be able to carry out its function before it expires. In either scenario, the microorganism would carry a suicide gene that is repressed when the microbe is at work and becomes active immediately thereafter (Curtiss, 1988; Molin et al., 1987).

Bacteria

The key to designing an effective suicide containment system rests on regulating gene expression from one of a variety of controllable promoters. They can be divided into two categories: those that function when a trigger is present and those that function until repressed (see Molin et al., 1993 for review). Systems that have been devised in the first category include the P_L promoter of phage lambda and a thermosensitive lambda repressor (Ahrenholz et al., 1994) and the *lac* promoter from *E. coli*. The P_L promoter is induced by raising the temperature to inactivate the lambda repressor; the *lac* promoter is activated by the chemical isopropyl-β-D-thiogalactopyranoside (IPTG) (Bej et al., 1988; Knudsen and Karlstrom, 1991; Knudsen et al., 1995). Although such systems work in the laboratory under controlled conditions, it is unrealistic to apply heat to fields or to irrigate fields with a chemical inducer such as IPTG. As a solution, Molin and colleagues (1993) suggest manipulating the regulated system such that growth of the cells in the laboratory leads to the synthesis of a compound that is toxic to the microbe. When cells are introduced into the environment, several generations of growth would be needed before the repressor would be diluted and the toxin synthesized. Because generation times in the wild can be just days or weeks long, the engineered cells should survive only long enough to achieve the goal. The approach is speculative, but it seems promising. For bioremediation, a possible approach involves repressing transcription from a promoter that functions only in the presence of the target substrate. Once the substrate is exhausted, transcription from the promoter ensures that suicide is induced (Contreras et al., 1991).

The systems that function when the activator is absent include the *trp* promoter from *E. coli*, which represses the synthesis of a toxic compound when tryptophan is present. If the microorganisms escape, their new environment would likely contain insufficient supplies of this amino acid, resulting in transcription of the toxin gene from the active *trp* promoter and synthesis of the toxin.

Two goals are paramount in the design of suicide systems. First, the gene product should extend beyond mere growth inhibition; the best candidates are killing functions whose targets are likely to be found in essentially all bacteria. Second, the toxicity of the gene product should be high, so that a high efficiency of killing is achieved at a low concentration. Therefore, putative killing functions should be assayed in various bacteria at several ranges of induction (Molin et al., 1993).

The strengths of suicide genes in bioconfinement are their high specificity and the variety of potential targets and activators. Molin and colleagues (1993) reviewed the major systems of suicide genes developed in bacteria. The *hok/sok* (host killing/suppression of killing) system originally was observed in bacterial plasmids (Gerdes et al., 1986). That system and others in the *gef* gene family consist of genes that encode for a toxic polypeptide that both attacks the cytoplasmic membrane and is the antidote to that polypeptide. If a cell spontaneously loses the plasmid (or if the gene inserted in the chromosome mutates) it dies because the leftover mRNA is translated into a toxic protein that degrades the bacterial membrane. One advantage of using this class of toxic factors is that their target, the cytoplasmic membrane, is similar in structure in many bacteria.

Other killing systems include nucleases, which target destruction of genetic material (e.g., Ahrenholz et al., 1994). Those systems are highly promising. Not only would they kill the engineered bacterium but they also destroy its DNA, which might otherwise be transferred from the dead organism to living cells via transformation. Genes that code for nucleases from *Serratia marcescens* and *Staphylococcus aureus* have been fused to an inducible *lac* promoter to create such killing systems. It is unclear to what extent the DNA repair systems in cells would make it difficult for the nucleases to degrade the DNA to the extent necessary to kill the cells. Lysis genes from bacterial viruses also have been considered as a source of killing genes. Those genes have been cloned and fused with regulated gene expression systems, with promising results (Molin et al., 1993). It is noteworthy that research in the development of suicide genes as a means of bioconfinement appears to have stopped about ten years ago for reasons that are not clear to the committee.

Fungi

The same methods that are being used to control the spread of transgenic bacteria are being applied to fungi. The object is to prevent fungal persistence and spread through the formation of various kinds of spores. If the spore is necessary for the fungus to execute its intended function, such as the infection of an insect, it is necessary to suppress sporulation after that goal is accomplished. Genes that inhibit sporulation could be put under the control of an inducible promoter and then engineered into the fungus. The spores would be treated with a chemical or environmental inducer before they are applied to the target pest.

Some fungi are being genetically engineered to contain genes that increase their virulence in specific insect pests (Hu and St. Leger, 2002; St. Leger et al., 1996). To prevent the creation of hypervirulent organisms, it has been proposed that the genes be flanked by antisense forms so as not to affect the virulence of the strain but to target genes in the recipient cells that might inadvertently receive the virulence genes. Such targets could involve reproduction, spore formation, and spore germination. To prevent vegetative spread of the mycelium, suicide genes could be engineered into cells under the control of an inducible promoter.

A major weakness of suicide genes in fungi and bacteria is the occurrence of mutations that prevent the system from operating. In large part, their usefulness has been demonstrated only in laboratory studies and it is not clear how they will function in the field. Some suicide systems are intriguing ideas that have yet to work even in the laboratory. Laboratory experiments show that killing by suicide gene systems is never absolute; a surviving subpopulation can continue to grow even in the presence of inducer (Molin et al., 1993). The survivors result from mutations in the killing gene, mutations in the expression system that inactivate the suicide function, or mutations in other parts of the cell that confer resistance to the action of the killing agent (Knudsen et al., 1995). Suicide systems also can be lost from the cell if they are located on a plasmid—the plasmid can be lost after transfer to only one daughter cell during cell division. One way to reduce the problem is through redundancy, provided by the use of two identical systems or the combining of different suicide systems (Jensen et al., 1993; Knudsen et al., 1995). Those efforts will lower, but not eliminate, the probability of mutations, resulting in resistance. Thus, suicide systems can reduce, but not eliminate, a genetically engineered population.

Viruses

To the committee's knowledge, suicide gene systems per se have not been applied in the production of genetically engineered viruses. However, viruses can mutate spontaneously into temperature-sensitive or *ts* mutants such that the mutant genotype cannot replicate at some temperatures. If a genetically engineered virus featured a *ts* mutation, that system could be harnessed in a fashion similar to the physical control of bacterial suicide genes. However, *ts* mutants typically are much less fit than are wild-type viruses, and using them might be more accurately described by the fitness reduction method known as phenotypic handicapping.

Failed or Inappropriate Methods of Microbial Bioconfinement

Chapters 3 and 4 describe the major bioconfinement methods used in higher organisms, and most of them are inappropriate for use in microbes. Because bacterial reproduction is strictly asexual, for instance, sterilization cannot be used to confine transgenic bacteria. Although some viruses can reproduce through reassorting chromosomal segments when multiple virus particles infect the same cell, the same viruses also can reproduce clonally. Thus, unlike most eukaryotes, they are not bound to obligate sexual reproduction. As a consequence, confinement of genetically engineered microbes must be limited to fitness reduction methods such as the induction of suicide genes or phenotypic handicapping.

Effectiveness of Methods at Different Temporal and Spatial Scales

Temporal scales could influence the effectiveness of bioconfinement in bacteria, fungi, and viruses. Although those microbes can grow rapidly under ideal laboratory conditions (up to one generation per hour), typically they grow much more slowly in nature (Madigan et al., 2003). Nutrients in the wild usually are limiting, and their scarcity can prevent microbes from achieving rapid exponential growth. In nature, bacteria often experience "feast or famine;" periods of rapid growth are interspersed with longer periods of retarded growth. In the transition, bacteria undergo dramatic changes in physiology and morphology, which adapt them to poor growth conditions. In periods of slow growth bacteria are much more resistant to environmental assault than are rapidly growing cells (Siegele and Kolter, 1992). The process of bacterial sporulation, in which bacteria enter a dormant, nonmetabolizing, highly resistant state, is an extreme example.

Fungi develop spores in the course of sexual or asexual reproduction. The spores, which are readily dispersed, are hardier than mycelia but not nearly as resistant to harsh environmental conditions as bacterial spores can be. Viruses are nonmetabolizing entities, so they do not have the luxury of regulating metabolism as do bacteria and fungi. Rather, as obligate parasites, viruses are at the mercy of their hosts (the biotic environment), and their ability to grow in adverse conditions (the abiotic environment) depends on host metabolism. Because the ideal growth conditions of host cells are likely to be separated in time, many viruses can infect their hosts latently until the hosts resume growth or active metabolism, which then would allow productive infection to occur again.

Fitness reduction methods are designed to hamper the reproductive potential of genetically engineered strains of bacteria, fungi, and viruses, placing them at a growth disadvantage relative to wild-type strains. These debilitated strains should fare no better and likely far worse than their wild-type counterparts in terms of ability to survive in the natural environment. However, insufficient field testing has been done in a variety of environments with different organisms and genotypes to confirm that expectation. A more important consideration for the effects of temporal scale on fitness reduction methods in any microbe is the possibility that the microbe will become latent (for viruses) or sporulate (for bacteria and fungi).

Effects of spatial scale on the confinement of bacteria and viruses are difficult to gauge; relatively little is known about dispersal of microbes in the wild. Most current data concern dispersal of pathogenic bacteria and viruses through physical processes (such as flow of water) or geographic movement of their host organisms (such as air travel by infected humans). Phenotypic handicapping and other fitness reduction methods are designed to reduce local survival of introduced bacteria, fungi, and viruses. Should those microbes become dispersed to distant locales, one might assume that the methods would be effective there as well. But this is not necessarily true, especially if the microbes are dispersed to different kinds of environments. Some viruses can inflict very different degrees of damage (becoming more virulent) when the host population is naïve to virus attack because of an absence of resistance alleles or antibodies in the host population (Bull, 1994; Taylor et al., 2001). Although highly speculative, a potential concern is that migration of genetically engineered bacteria, fungi, and viruses to new places could release them from phenotypic handicapping and from other mechanisms designed to hinder fitness.

Because the effects of sporulation and germination traditionally have not been evaluated in field tests, it would be wise to avoid, if possible, genetic modification and release of sporulating microbes as a way to minimize the risk of long-term survival and dispersal and allay the fear that those strains would transfer their genetic material to local populations.

Ecosystem and Population Effects

The committee has identified phenotypic handicapping and suicide systems as the primary methods that could be used for bioconfinement of bacteria, fungi, and viruses. Although more field data are needed, it is unlikely that the methods themselves would damage ecosystems and natural populations. However, because they have been studied only under laboratory conditions, three conceivable consequences can be foreseen that echo the motivations for bioconfinement outlined above. All relate to the environmental consequences of failure under natural conditions.

Although there are no supporting data, the possibility exists that release of transgenic bacteria, fungi, and viruses can damage indigenous microbial populations. For instance, although the effects of introduced viruses on intended hosts can be gauged accurately through laboratory experiments, the introduced viruses could attack nontarget hosts that are ill prepared to defend themselves. Similarly, survival of bacteria in the wild could be underestimated from laboratory results. The second concern involves displacement of resident species. Indigenous viruses could be less virulent than introduced ones, creating the opportunity for engineered viruses to severely reduce the population size of local hosts. Finally, genes from introduced bacteria, fungi, and viruses could be transferred horizontally into resident species. Because those genes could have unanticipated effects when they migrate to new backgrounds, they could reduce fitness at one or more loci through negative epistasis. Although selection should act to remove the introduced genes from the local gene pool, it could take a long time for dangerous alleles to be completely removed as a result of weakened selection as the genes become rarer in the population (Hartl and Clarke, 1997). The committee believes that the ecological consequences of using fitness reduction methods, such as phenotypic handicapping and suicide systems in genetically engineered microbes, are likely to be minimal, because those methods are designed to employ genotypes that are competitively inferior indigenous strains. However, because the methods have not been evaluated in the field, it is not possible to state with certainty that they will have the desired effect in confinement.

Monitoring, Detection, and Culling: Needs, Feasibility, and Realities

The frequency of genetically engineered microbes in natural environments can be estimated rather straightforwardly if natural populations are extensively sampled and screened with modern molecular techniques, especially if the engineered organisms contain easily detected phenotypic markers, for example, that are visible on a selective agar medium. However, it is virtually impossible to completely eliminate specific genotypes in natural populations of microbes (Salyers and Whitt, 2002). This needs to be considered when deciding whether a genetically engineered microbe should be released into the environment.

Microalgae

Although this chapter focuses on bioconfinement of transgenic bacteria, fungi, and viruses, the possibility of bioconfinement also should be evaluated for genetically engineered microalgae.

Microalgae have already been successfully engineered (see review by Minocha, 2003). The best results have been obtained with *Chlamydomonas reinhardtii*, which has long served as a model system for physiological and molecular studies (e.g, Cerruti et al., 1997; Dunahay, 1993). In particular, genetic engineering of *C. reinhardtii* could be useful for bioremediation of heavy-metal pollution, a pervasive environmental problem because trace metals cannot be decomposed but must be sequestered from the environment. Cai and colleagues (1999) demonstrated that the trace-metal-binding properties of *Chlamydomonas* can be enhanced in transgenic genotypes that express a foreign-metal-binding protein, without slowing their growth rate relative to wild-type cells. In addition, stable nuclear transformation has been achieved in the colonial green alga *Volvox carteri* (Hallman and Sumper, 1994; Schiedlmeier et al., 1994). A few diatoms also have been successfully transformed, including the widely studied model system *Phaeodactylum tricornutum* (e.g., Apt et al., 1996). This is promising because diatoms have commercial uses as feed in aquaculture and as potential sources of useful pharmaceuticals.

Most commercial-scale cultivation of microalgae is performed in large, open outdoor ponds. Zaslavskaia and colleagues (2001) identified several disadvantages of this approach, including invasion of ponds by contaminants and reduction in biomass production resulting from seasonal and diurnal variations in temperature and light. Thus, improved efficiency and reduced cost of micro-algal biomass production could be achieved if the microbes were engineered to grow as heterotrophs in conventional microbial fermenters (in the absence of light). Zaslavskaia and colleagues (2001) introduced a gene that encodes a glucose transporter into the obligate photosynthetic microalga *P. tricornutum*, allowing the diatom to thrive on glucose in the absence of light. The approach seems promising because fermentation technology eliminates contamination by microbes, which is an important criterion for maintaining food industry standards.

Microalgae are biologically similar to bacteria (especially photosynthetic bacteria) that grow in aquatic environments, and they have similar mechanisms for horizontal gene transfer. Therefore, the same consequences and concerns would apply to their bioconfinement. However, most transgenic microalgae have been cultivated in closed-system indoor tanks and are not intended for release into natural environments. Because of their similarity to bacteria, phenotypic handicapping and suicide systems should provide effective bioconfinement if necessary. The committee did not find any reports in the literature of efforts to test the feasibility of those methods in microalgae.

6

Biological and Operational Considerations for Bioconfinement

This chapter summarizes and analyzes what has been presented in the foregoing chapters. First, the biological opportunities and constraints for confinement are reviewed, with special emphasis on bioconfinement. Next, the operational implications of confinement are considered. Then, confinement failure and its mitigation are discussed. Finally, there is a look to the future and the need to explore unanswered questions and promote research that will build better avenues for the confinement of genetically engineered organisms (GEOs).

WHAT BIOLOGY TELLS US ABOUT CONFINEMENT AND BIOCONFINEMENT

As explained in Chapters 3–5, a wide array of bioconfinement measures has been proposed for limiting the movement of transgenes. Some of them are hypothetical, some have been examined in the laboratory, and a few take advantage of well-known biological phenomena. All of them share some features. Each method has strengths and weaknesses, and all vary in efficacy depending on circumstances. No one method will achieve 100% confinement in the real world. Straightforward conclusions follow from the observations presented in the preceding chapters.

Case-by-Case Evaluation

GEOs represent a heterogeneous class with regard to biosafety. Some present minimal risk, others moderate risk, and yet others considerable risk. As noted in Chapter 2, the decision of whether and how to confine a GEO depends on factors that range from the phenotype associated with the transgene to the environment into which the organism would be released. Confinement options will vary with the precise species chosen for transformation because there are so many differences in size, genetics, ecology, and dispersal biology. In some cases where confinement is necessary, physical and physicochemical confinement options will suffice; in others, biological confinement might be necessary. Clearly, there is no universal option, and case-by-case evaluation is a necessity.

Finding 1. The efficacy of bioconfinement will depend on the organism, the environment, and the temporal and spatial scales over which the organism is introduced.

Finding 2. In many cases GEOs will not require bioconfinement.

Recommendation 1. Evaluation of the need for bioconfinement should be considered for each GEO separately.

Early Evaluation

The evaluation of whether and how to confine a GEO cannot be an afterthought in the process of development of a transgenic organism. Making biosafety a primary goal from the start of any project will be a more effective and efficient way to prevent safety failures and it will increase commercial investment ratings and reduce financial risks posed by possible liability claims and loss of consumer confidence (Kapuscinski et al., 2003). If biosafety considerations are delayed until after a product is developed, the need to receive a return on the investment made to create that product could cloud the judgment of those who determine whether and how it should be used. Similar considerations (including reducing liability and avoiding public relations problems) make it preferable for noncommercial GEO developers, such as universities or international research centers, to make biosafety a primary goal at the outset.

Dispersal biology and the opportunities for the unintentional movement of transgenes must be considered as part of the process of finding the best organism to modify to create a product. For example, the evaluation should consider whether the organism is to be released near or distant from other organisms of the same species. Early evaluation permits the consideration and comparison of simpler, traditional confinement techniques alongside the more complex, and sometimes more expensive, bioconfinement options. The constant and iterative evaluation of confinement options during the development of a GEO should optimize both the efficacy and the cost effectiveness of the confinement options once they are deployed. Hurried consideration just before the deployment of a GEO is apt to create a makeshift and expensive plan that might work better in theory than in practice.

Recommendation 2. The need for bioconfinement should be considered early in the development of a GEO or its products.

Redundancy

Because methods can fail, a single confinement method will not necessarily prevent transgene escape. Therefore, it is sometimes necessary to employ more than a single method. In many technological applications, the principle of redundancy reduces the occurrence of predictable hazards while achieving the benefits of technological application. Redundancy involves applying two or more safety measures to product design and use, each with fundamentally different strengths and vulnerabilities, so that the failure of one safety measure is counterbalanced by the integrity of another. In other cases, it may be possible to combine two barriers of the same type but whose failures would be independent events, such that a failure of one barrier does not trigger a failure of the other.

This does not necessarily require using different bioconfinement methods, as long as the measures are independent. By mixing confinement measures with different vulnerabilities, the chances improve that failure of one safety measure will not breach the target level of confinement. When choosing redundant confinement techniques (including bioconfinement), measures should be chosen to compensate for each other's weaknesses.

In many cases, this will involve application of an appropriate mix of biological, physical, and physicochemical confinement measures tailored to the GEO in question (Agricultural Biotechnology Research Advisory Committee, 1995; Kapuscinski, 2001; Scientists' Working Group on Biosafety, 1998). For example, the U.S. Department of Agriculture (USDA) has developed requirements for growing transgenic corn for pharmaceutical and industrial chemical production that mandate spatial and temporal isolation from corn grown for other uses (Federal Register, 2003). One feasible application of the principle of redundancy in aquaculture would be to combine physical barriers, such as floating cages, with bioconfinement consisting of the use of all-female lines of sterile, triploid fish.

Finding 3. It is unlikely that 100% confinement will be achieved by a single method.

Finding 4. Redundancy in confinement methodology decreases the probability of failing to attain the desired confinement level.

Experimental Information on Efficacy

The discussion of redundancy implies that some information should be available on how well a confinement method works. The effectiveness of many confinement methods, particularly bioconfinement methods, will depend on the genotype and the environment. Thus, the efficacy of the planned combination of confinement methods should be tested in representative genotypes under development to ensure that the plan is effective. Also, the planned combination of confinement methods should be tested in every environment in which it is anticipated that a GEO will be released— including any environment that the GEO could be foreseen to occupy. For example, if strict confinement is desired for a corn genotype that is to be grown from seed, it is important to test the efficacy of the confinement technique in all environments to which that seed might accidentally be dispersed.

Likewise, before field release, the reproductive biology of the novel genotype should be measured relative to its progenitor to evaluate changes that might affect its rate of gamete and progeny production and their dispersal. Studies have shown that some transgenes could allow wild or weedy relatives of the crop to be more successful and that other transgenes will not (Burke and Rieseberg, 2003; Snow et al., 2003). Although new genotypes generally do not have reproductive phenotypes that are different from those of their parents, any changes that occur can be dramatic and have important consequences. For example, hybridization between non-GEO sugar beets and wild sea beets introduced an allele into a crop that increased its rate of premature flowering, and the crop became a noxious weed (Boudry et al., 1993, 1994; Viard et al., 2002). Changes in reproductive biology might not be an anticipated phenotype associated with a novel genotype. Pleiotropy—the unanticipated phenotypic effects of a single allele—is not rare. Beet–Swiss chard hybrids with a transgene construct for virus resistance showed a *decreased* rate of premature flowering relative to nontransgenic control plants (Bartsch et al., 2001).

Recommendation 3. Confinement techniques should be tested experimentally, separately and in combination, in a variety of appropriate environments, and in representative genotypes under development before they are put into application.

Recommendation 4. To evaluate changes in reproductive biology, the novel genotype should be compared with that of its progenitor before field release. For long-lived species, such as trees, it may be necessary to begin field tests before such comparisons are possible, with a realistic plan to mitigate any unexpected and dramatic increase in reproduction.

Changes of Efficacy with Scale

Typically, precommercial evaluation of GEOs starts at a small scale and then is often expanded to larger scales before release. Even with the largest precommercial field trials involving up to 100 sites and 1000 acres (or less) per site over a two- to three-year period, the scale of these may be dwarfed by the regional or continental scale at which these GEOs may be produced. It is well known that many environmental concerns cannot be addressed prior to commercialization. An example of a response to these concerns is the monitoring requirements for *Bt* resistance in target insects (NRC, 2002a). Similarly, the spatial or temporal scale of a field release can influence the potential for confinement failure. The appropriate confinement option will depend on scale. Under a very limited field release—a tenth of an acre or over a few hours—one or two methods of confinement might suffice. However, the same genotypes released over 100 acres or for many years could require several methods to obtain the same level of confinement. Alternatively, field release might not be a safe option on a large scale or for a long period. If possible, empirical data (experimental or otherwise) should be used to determine whether the confinement plan is adequate for the anticipated scale of field release.

Recommendation 5. Bioconfinement techniques should be assessed with reference to the temporal and spatial scales of field release.

How Much Bioconfinement is Enough?

The foregoing sections suggest the need to define "adequate level of bioconfinement" early on. This requires an evaluation of failures and their consequences under worst-case scenarios. It also requires an assumption that escaped genes have the opportunity to multiply. In some cases, the escape of 10 individuals per year into the ambient environment might not be a problem; in other cases, 10 would be too many.

Recommendation 6. An adequate level of bioconfinement should be defined early in the development of a GEO, after considering worst-case scenarios and the probability of their occurrence.

Unacceptability of Some Methods under Some Circumstances

Some bioconfinement methods will be unacceptable under some circumstances. Apomictic seed production by absolutely sterile male plants could be a multiple benefit in ensuring true-to-type seed without an opportunity for transgene escape by pollen. Combined with multiple confinement methods, the use of apomixis could be acceptable. However, apomictic organisms with some male fertility that are released close to wild relatives pose an opportunity for the transgenic genome to sweep through the wild population, replacing it with a clonal transgenic lineage (van Dijk and van Damme, 2000; see Chapter 3). In that case, the use of apomixis should be rejected as a confinement method.

Finding 5. Some bioconfinement methods are unacceptable under some circumstances.

Options Based on Technology and Gene-Specific Compounds

Bioconfinement methods that are based on transgenic technology have received considerable recent attention (e.g., Daniell, 2002), and this committee also has identified the potential of bioconfinement by external administration of gene-specific compounds (Chapter 4). Although those methods hold great promise, none has been tested in an array of organisms and for a variety of environments. Indeed, some methods are still theoretical. Even those transgenic bioconfinement methods that have been created have yet to be tested adequately in a single organism under a variety of field conditions. Statistically adequate experiments still are necessary to measure their efficacy. Transgenic and gene-specific bioconfinement technology still is in its infancy and has not yet been proven as effective as have nontransgenic confinement methods that already are in use.

Finding 6. Many types of bioconfinement are still in the early stages of development, especially those based on transgenic methods and gene-specific compounds.

EXECUTION OF CONFINEMENT

The foregoing considerations suggest that the field release of a GEO constrained by confinement should follow a straightforward pathway:

Decision Making

Once the phenotype of the organism has been identified, its biosafety must be appraised: What risks does it pose? What would be the worst possible scenarios created by those risks? How are those risks balanced by anticipated benefits? Is some confinement necessary? If so, how much? Are the risks large enough to warrant the use of a different organism or abandoning the project altogether? What are the possible opportunities for the escape of the gene or the organism, including human error in handling living propagules or gametes? What confinement methods are available for this organism? What is the potential for spread of the GEO if it escapes? Given what is known about the methods, the organism, the novel phenotype, and the spatial and temporal scale of anticipated field release, which combination of methods—physical, physicochemical, and biological—should suffice to obtain sufficient confinement to make the risk acceptable?

Research

Assuming that the decision is made to proceed with a project and that confinement is warranted, experiments designed to answer the questions above should be conducted. For example, the efficacy of the proposed combination of confinement methods should be tested in the field before their use with genotypes that are as similar as possible to the novel genotype in question. The proposed combination of confinement methods also should be tested in an appropriate range of environments to which the new genotype will be released or to which it might escape. Similarly, the reproductive biology of the novel genotype should be compared with that of its progenitor before field release to evaluate changes in reproductive biology.

Integrated Confinement System

If the tests of the proposed confinement technique suggest that it will be successful, it will still be necessary to establish an integrated confinement system (ICS) for the deployment of the organism to ensure confinement efficacy, especially as the new genotype is spatially and temporally deployed. ICS is a systematic approach to the design, development, execution, and monitoring of the confinement of a specific GEO. This recommendation is in keeping with system safety management as it is widely practiced in the management of many modern technologies (Roland and Moriarity, 1990). System safety management is a forward-looking, comprehensive, long-term approach that ensures that systems and techniques have safety designed in from the outset (McIntyre, 2002). Necessary elements of ICS include the following:

- Commitment to confinement by top management
- Establishment of a written plan for redundant confinement measures

to be implemented, including documentation, monitoring, and remediation (in case of failed confinement)

- Training of employees
- Dedication of permanent staff to maintain continuity
- Use of standard operating procedures for implementing redundant confinement measures
    - Use of good management practices for applying confinement measures to pharmaceutical-producing GEOs or the equivalent
    - Periodic audits by an independent entity to ensure that all elements are in place and working well
    - Periodic internal review and adjustment to permit adaptive management of the system in light of lessons learned
- Reporting to an appropriate regulatory body

For an ICS to work, it should be supported by a rigorous and comprehensive regulatory regime that is empowered with inspection and enforcement.

Recommendation 7. An integrated confinement system approach should be used for GEOs that warrant confinement.

Monitoring and Detection Technology

The efficacy of the confinement system must be monitored constantly. Several detection techniques are available to determine whether transgenes move to organisms or environments as the result of confinement failure. Some are associated with portions of the transgenic construct. The creation of a transgenic organism usually involves a selectable marker, such as resistance to a specific antibiotic or herbicide. Because the chosen trait is unlikely to be present in nonengineered members of the species, it can serve as a reliable marker. Likewise, the creation of a transgenic organism sometimes involves inserting a reporter gene to confirm that the promoter is working effectively. For instance, the so-called GUS construct is a reporter gene that creates a blue color when cells are soaked in the appropriate solution (Jefferson et al., 1987). Although still at the research level, product developers may be able to use the Cre/lox-mediated recombination technology (see Chapter 3) in the future to remove the unwanted selectable marker genes. Finally, there are methods for testing directly for the genotype, product, or phenotype of the transgene. For example, a standard tool for amplifying a specific DNA segment (polymerase chain reaction, or PCR) facilitates testing for the presence of specific transgene constructs in the genotype of an organism. A standard testing method for detection of a specific protein (enzyme-linked immunosorbent assay, or ELISA) is available for testing for the presence of *Bt* protein. Herbicide resistance can be tested by direct application of the appropriate herbicide at the appropriate concentration (e.g., Lefol et al., 1996). The committee notes that our ability to detect transgenes with PCR and other devices may currently exceed our ability to characterize the risk or consequences from such transgene contamination. As noted earlier in the report, an adequate and appropriate characterization of such events will almost always be on a case-by-case basis—depending, for example, on the transgene, its function, the environment where the contamination occurs, the species carrying the transgene, and the number of GEOs involved.

In the future, organisms might be transformed with additional constructs for the purposes of monitoring them. The addition of a gene derived from jellyfish that expresses green fluorescent protein has been used to monitor insects released for biological control (Staten et al., 2001), and it has been proposed for tracking transgenic plants (Leffel et al., 1997). Likewise, insertion of DNA sequences that can be used as "bio-barcodes™" to identify specific transgene constructs has been proposed (Gressel, 2002). Ideally, the development of monitoring methods that can identify escapes through remote sensing and that use Geographic Information System technology would make monitoring more feasible.

Monitoring of bioconfinement will not be a simple matter. It will involve looking for what will often be a rare event over a potentially large area. Under such circumstances, sampling becomes a challenge (Marvier et al., 1999). The seeds, eggs, pollen, sperm, spores, or other dispersal propagules of many organisms often are too small to collect or analyze in any statistically meaningful way. The expense and effort of adequate monitoring could outweigh the perceived benefits of introducing a GEO to the field. Even with the best and most thorough monitoring scheme, some events will be missed, and, given enough genotypes over enough time, some fraction of those events will have negative consequences.

However, even failures of monitoring can offer benefits. A monitoring failure can be used as an example for developing better confinement and monitoring techniques. Catching a mistake too late may still allow the identification of the source of the product. In the case of realized harm, it can be used to assign responsibility.

Nonetheless, monitoring should be seen as a complement to confinement, not as a replacement for it. That is, the act of monitoring should not result in complacency about the possibility of escape. Effective confinement and adequate monitoring are often easier to manage than eradicating a reproducing organism once it has reached critical numbers (Simberloff, 2003).

Recommendation 8. Easily identifiable markers, sampling strategies, and methods should be developed to facilitate environmental monitoring of GEOs.

Eradication or Control of Escaped Organisms

It can be worthwhile to attempt to eradicate or control escaped GEOs or transgenes. If individuals can be identified easily and the escape is localized, eradication can be possible, depending on the organism. If the escaped organisms do not appear likely to cause the harm originally anticipated, it can be worth considering whether control is necessary at all, especially if control will be difficult. If detection has come too late, however, and if the organisms are creating problems and are too widespread for eradication, control is the only option.

Increasing the Efficacy of Confinement

Three issues that can significantly affect the efficacy of bioconfinement measures are not directly related to natural science: transparency and public participation, compliance, and international considerations.

Transparency and Public Participation

The public's right to information—often called transparency—and to participate in decision making, are fundamental principles of democracy. Each right complements the other, and each can improve the effectiveness of confinement. For example, public participation can bring otherwise unknown information to the decision-making process. Transparency can increase acceptance of bioconfinement measures (and of the GEOs being confined) by building public trust in the decision-making process. Transparency and public participation also can improve the quality of decisions about GEOs and confinement in terms of protecting human health and the environment. This is true at various stages of decision making about GEOs and confinement. Confinement considerations should come into play at a number of stages in the "life cycle" of a GEO, including research to develop and characterize the genetic and phenotypic traits of a GEO, risk analysis and risk reduction, field testing, commercialization, large-scale production, processing, transportation (domestic or international), and disposal.

The analysis associated with selection of confinement methods for GEOs—including the decision to proceed or not—would benefit from having a public component. Public participation can identify hazards, raise important questions, and provide information about specific conditions that can lead to more realistic assumptions (NRC, 1996; Hails and

Kinderlerer, 2003). Members of the public, who often will not be scientific experts or otherwise involved in the field of genetic engineering, can offer information that is indispensable to the clear understanding of social values and other factors that affect the significance of potential effects of a confinement failure (Chapter 2; NRC, 2002a). Transparency also is important to the assessment of environmental or health effects. Transparency about novel GEOs and their confinement also could yield significant benefits in the face of failure. In some cases, when human health or the environment could be at risk, transparency would increase the likelihood that failure can be averted or mitigated early enough to prevent harm.

The committee emphasizes that, to its knowledge, no significant harm to health or the environment has resulted from GEO confinement failure. Nonetheless, the StarLink and Prodigene incidents (described in Chapters 1 and 2), are examples of failures in a system that is intended to maintain safety. This is the same system that the American public expects to ensure food safety and environmental protection as a growing array of new GEOs comes into production. Greater challenges for risk assessment and management (Chapters 3–5) will be faced as the probability of a confinement failure increases with use of GEOs on larger spatial and temporal scales and with their growing application to produce an increasingly wide array of products. A lack of transparency could increase the likelihood of failure of confinement and exacerbate its consequences.

The committee believes that, because of the fundamental need for transparency and public participation, the close connection between them, the need to safeguard the environment, and the desirability for increased credibility with respect to GEOs and their confinement, close cases should be called in favor of transparency. When the need for intellectual property protection of a bioconfinement method arises it will influence how transparency is maintained; however, transparency should remain a priority. The committee also believes that appropriate transparency and public participation should be promoted in designing and implementing the systematic approach to confinement—the ICS described earlier. The appropriate degree and nature of transparency and public participation could vary at the different points in the system.

Recommendation 9. Transparency and public participation should be important components in developing and implementing the most appropriate bioconfinement techniques and approaches.

Compliance

Compliance is critical to the success of confinement. If the method in question is not followed, bioconfinement will fail—regardless of its theoretical efficacy. The committee considered a few of the many factors that can influence compliance: the nature of bioconfinement methods and the state of verification and monitoring technology, human error, natural events, the cost, and increases in spatial and temporal scale.

Confinement Methods, Verification, and Monitoring

Compliance with a chosen or prescribed confinement measure is affected by how difficult that measure is to apply. Compliance can be expected to increase—other things being equal—as the ease of applying the confinement measure increases. The efficacy of confinement also would vary with the human processes involved, because of human error, discussed below, and because a properly designed management process can improve implementation of prerelease verification and postrelease monitoring.

A related factor is the difficulty for those who produce and use GEOs, as well as for regulators, of verifying and monitoring confinement efficacy. This could depend on the characteristics of the GEO or of the confinement method and on the technology available to test for the presence, or measure the effectiveness, of confinement methods. Verification and monitoring technology are discussed earlier in this chapter.

Some bioconfinement methods are more amenable than others to verification and monitoring. It is easier to verify a bioconfinement technique that has an obvious physical manifestation, such as one that involves a visually identifiable phenotype, than it is to verify a bioconfinement technique that does not.

Finding 7. The efficacy of bioconfinement will vary with the human processes involved in applying the methods, with the characteristics of the GEO, and with the confinement method itself.

Human Error

Humans make mistakes, and experience with GEO confinement bears this out. In the 1999–2000 StarLink situation (see Box 2-1, Chapter 2), corn for human consumption was contaminated by a genetically engineered variety approved only as an animal feed (Taylor and Tick, 2003). The commingling probably occurred because the U.S. commodities system does not keep bulk grain separated. There was considerable speculation as to how and why the varieties were mixed, but there was no doubt that human error was a major cause.

The committee recognizes the difficulty of predicting when and where human error will occur, particularly for bioconfinement, for which there is no history of mishaps from which to try to generalize or predict. The committee nevertheless believes that the probability of human error should be considered, for instance, by drawing on methods of system safety (Roland and Moriarity, 1990) or organizational analysis, which could be refined as data about confinement accumulate over time.

A peculiar form of human error, which has its roots in kindness, also could affect confinement. For example, many goldfish owners do not wish to kill pets they no longer want, and instead might release the fish into bodies of water or flush them down a drain. This is an error in the sense that the actor presumably is ignorant of the possible consequences of the action, which are not always benign. Red-eared slider turtles (*Trachemys scripta elegans*) and snakehead fish (*Channa marulius, C. argus, C. striata, C. micropeltes*) have been introduced throughout their nonindigenous range through pet releases (Mayell, 2002; USFWS, 2002; USGS, 2002). Similar actions could affect confinement of genetically engineered animals that have reached the end of their "production lives," yet remain healthy.

The omnipresent risk of human error was an important factor underlying the committee's conclusions that the implementation of confinement methods should be systematic and integrated. Redundancy in confinement methodologies is essential.

The committee is aware that intentional human actions, such as bioterrorism or unethical business practices, might result in a failure of bioconfinement and release of GEOs into the environment. These topics are beyond the scope of this study; another NRC study has considered one aspect of this issue (NRC, 2003b).

Recommendation 10. The possibility of human error should be taken into account as a factor when determining bioconfinement methods and evaluating their efficacy.

Natural Events

Compliance also can be affected by natural events. A hurricane, tornado, or tsunami can wreak havoc with physical confinement, for example, by destroying fish cages. Similarly, natural vectors such as insects, rodents, and birds that carry seeds can affect dispersal. If a bioconfinement system is dependent on physical confinement, a natural disaster could expose the organism to an environment where the bioconfinement technology would no longer function optimally.

Cost of Compliance

One would expect that—all other things being equal—compliance will increase as the cost of a prescribed bioconfinement regime decreases. The committee is aware, however, that cost can vary significantly. Switching from a well-known and genetically well-characterized crop, such as corn, to a less well understood plant for chemical production could add greatly to the cost of its confinement. Many bioconfinement techniques are expensive simply because they are untested or are still in the early stages of development. As new techniques that emerge from laboratories and field trials are put into use, the cost of implementation should change. The committee was thus not able to determine how the cost of compliance might favor specific bioconfinement techniques.

Private Litigation

Business entities have an incentive to comply with bioconfinement to reduce their risk of liability from private litigation arising from damage to human health or the environment. Private actions alleging liability for damage caused by the escape of a GEO (a confinement failure) can be brought under state tort or nuisance law, although few cases have been filed (Chapter 2). The strength of any disincentive effect will depend not only on the extent of liability provided by relevant tort and nuisance laws but on the interplay with intellectual property law. Under some circumstances, confinement failure can lead to allegations that intellectual property rights associated with GEOs have been violated by a third party. Those cases will thus help set precedent in cases that pertain to bioconfinement failure. Compliance also can be affected by private suits that are brought to enforce federal laws or to challenge the way federal agencies respond to citizen petitions regarding GEO confinement.

Increases in Spatial and Temporal Scale

Increases in the spatial and temporal scale of GEO production and the use of bioconfinement techniques could affect the incidence of compliance. This is discussed in detail in previous chapters and earlier in this chapter.

INTERNATIONAL ASPECTS

GEOs have several significant international dimensions that are relevant to confinement, as described in Chapter 1. The biotechnology industry is international, and development, testing, and use of GEOs actively (and increasingly) occurs throughout the world. GEOs are traded internationally. GEOs also can move across national boundaries by a wide range of mechanisms. Therefore, no single country can regulate all of the confinement issues that could affect its citizens, its economy, or its environment.

The international obligations of government must be linked to a case-by-case analysis of the GEO.

When confinement fails, GEOs can move from one nation to another. Addressing confinement here in the U.S. thus requires considering efficacy, concerns, and consequences not only in this country but in other countries to or from which GEOs are likely to move. International mechanisms and regimes that apply to such movement also are of concern. The U.S. thus has an interest in international cooperation on appropriate bilateral or multi-lateral regulatory regimes and in appropriate activities for standardizing regulatory approaches in countries throughout the world.

Recommendation 11. Regulators should consider the potential effects that a failure of confinement could have on other nations, as well as how foreign confinement failures could affect the United States.

Recommendation 12. International cooperation should be pursued to adequately manage confinement of GEOs.

BIOCONFINEMENT FAILURE

Bioconfinement measures occasionally will fail, for example, because of human error or because of an unpredicted response in the GEO. Poten-tial problems can be addressed at two different times—before and after a failure occurs. The committee focused on preventive actions that might be taken to prevent escape of GEOs and their genes and to questions that would be most important: What, if any, bioconfinement measures (possibly used in concert with other confinement measures) should be used to pro-vide the desired confinement? And, if that amount of prevention is not achievable, should the genetic engineering of the organism proceed at all?

The optimal choice of bioconfinement method for any particular situa-tion will be unique to that case. Chapter 2 discusses the points to consider: the desired or mandated level of protection; the organism; the novel trait; the available confinement techniques, biological and otherwise; the relevant genomic, physical, and biotic environment; behavioral factors; social values; the resources potentially available for prevention or remediation of bio-confinement failure; and the competing demands for those resources. In addition, the applicable regulatory regime could impose requirements or constraints regarding what confinement techniques may be used.

Decisions about confinement—and remediation if confinement fails—depend on the judgments, values, instincts, and skills of the people and organizations involved in decision making, as well as on the political situa-tion. Chapter 2 presents an approach that attempts to provide guidance to decision makers based on a rough analysis of the severity of consequence and the probability of occurrence. That and other chapters present several rules of thumb that could assist those who must make judgments.

As indicated above, the need for whether confinement is necessary should be considered, and an adequate level of confinement should be defined, early in the development of a GEO (Recommendations 2 and 6). This reflects the fact that it is essential to consider possible preventive action, including the use of biological and other confinement measures. With respect to harm to human health and the environment generally, it is well recognized that prevention typically is less expensive and more effective than post-failure remedial action, and that some consequences (e.g., death of a human, extinction of a species, and destruction of a major ecosystem) cannot be undone at all. Indeed, the choice of what confinement technique or techniques to use should be made very early in the process of developing a GEO as part of a broader analysis of possible preventive actions: confinement may be precluded if not undertaken early, and that analysis may determine that the desired level of protection cannot be attained through the use of biological and other confinement measures. Thus the proposed GEO may not be developed at all.

LOOKING TO THE FUTURE: STRATEGIC PUBLIC INVESTMENT IN BIOCONFINEMENT RESEARCH

The need for continued—and increased—public support of agricultural research has been articulated in previous National Research Council reports (NRC, 1989b; 2000; 2003a). One NRC report (NRC, 2002c) defines publicly funded agricultural research as any agricultural research performed with financial or material support from the public sector. For agricultural research in general, the reasons given for public support include

• To improve human health and well-being through advances that lead to higher quality and nutritional value in the food supply and greater food safety
• To sustain the quality and productivity of natural resources
• To preserve biological resources that are the endowment for future generations

Publicly funded research on bioconfinement methods is needed for all of these reasons.

The institutions that conduct and fund public agricultural research have been widening their agendas to support broad public policy goals. Environmental issues, sustainable production systems, and resource conservation are among the new emphases (NRC, 2002c; 2003a). That shift is occurring simultaneously with an increase in industry-funded agricultural research. In the past, publicly funded basic *and* applied research—mainly at the nation's land grant colleges and universities and in USDA and state laboratories—focused on productivity. The result was the underpinnings of the large agricultural output we enjoy today. Although strong public-sector crop and animal-breeding programs continue, over the past 25 years much of the productivity research has moved to the private sector, as is especially apparent in applied research in biotechnology. Of necessity, industrial research is primarily market-driven, whereas publicly funded research need not be.

Long-range and non-market-driven publicly funded research can help ensure the continuation of the fundamental biological discoveries that will lead to innovative bioconfinement methods not envisioned today. Publicly funded research also should lead to new ways to assess the risks of various bioconfinement—and other confinement—methods, and to new ways to monitor confinement. Finally, publicly-funded research on bioconfinement will help train professionals who will manage this powerful technology.

In addition to these broad reasons for publicly funded research on bioconfinement, **the committee recommends support for additional scientific research that**

- **characterizes ecological risks and consequences and develops methods and protocols for assessing the environmental effects of confinement failure (Recommendation 13).** More data are needed on the nature of potential ecological effects: their probability, their severity, and the potential for remedial action should confinement fail. Those research needs also were identified in recent reports (NRC, 2002a; 2002b) that noted the need for developing deeper theoretical and empirical understanding of the kinds of environmental effects that could result from transgene movement and the conditions under which such effects would be likely to occur. Many novel transgenic organisms are likely to be developed, and it will be useful to fund research to identify and investigate environmental hazards associated with a range of transgenic plants, animals, and microbes (NRC, 2002a; 2002b).
- **develops reliable, safe, and environmentally sound bioconfinement (Recommendation 14).** Clearly, this is especially important for GEOs that have a high potential for escape, such as perennial plants (turfgrasses and trees), aquatic organisms, insects, microbes, and viruses. The need for continued research on bioconfinement of genetically engineered crops was noted in an earlier report of the National Research Council (2002a).

This committee suggests that a special case can be made for research aimed at identifying and developing new hosts for transgenes involved in the production of chemicals and pharmaceuticals. Those hosts should have features that prevent them from threatening the environment and its ecology, biodiversity, and the food and feed supply. For example, tropical plant hosts that have no known temperate relatives and an inability to overwinter in temperate regions might be grown in those regions in the summer. Those plants also could be grown in greenhouses year-round, and their ability to escape would be limited by winter temperatures. The research could involve everything from basic investigations of the biology and genetics of the new hosts to their cultivation, harvest, and processing. By its nature this would be long-range and high-risk work, and it is therefore unlikely to be attractive to private industry. The committee notes that such research would be expected to lead to the development of new niche crops and to new industries to process the products. Furthermore, **the committee also recommends support for scientific research that**

* **identifies and develops methods and protocols that assess the efficacy of bioconfinement (Recommendation 15).** It is important to know how a given method performs in various environments and across different spatial and temporal scales. In this context—and for environmental and safety studies as well—new methods and approaches are needed for monitoring confinement. Easily identifiable markers for GEOs would be particularly useful.
* **identifies economic, legal, ethical, and social factors that might influence the application of particular techniques, as well as their regulation (Recommendation 16).** Evaluating confinement requires a multidisciplinary approach that includes the natural and social sciences. No collection of such expertise exists in the U.S.—or anywhere else—to the committee's knowledge, and the use of social science information in this area is particularly weak. Specific issues on which research could shed useful light include social factors that affect the significance of potential hazards; behavioral patterns of those who grow or use GEOs with respect to the willingness or ability to apply appropriate bioconfinement techniques; ways to reduce human error and instill a strong confinement ethic in those engaged in bioconfinement; and ways the federal regulatory system could be simplified, strengthened, and made more credible.
* **develops a better understanding of the dispersal biology of organisms targeted for genetic engineering and release, where sufficient information does not exist or where questions have arisen (Recommendation 17).** In particular, the following issues have been neglected: seed dispersal patterns, the significance of rare long-distance dispersal, the population genetic impacts of repeated and unilateral migration, the relative fitness effects of transgenes in introgressed organisms, improved verification technology, and the theoretical and empirical bases of monitoring.
* **develops a better understanding of invasion biology (Recommendation 18).** In particular, the ability to predict invasiveness is weak. Research

should address what ecophysiological changes or other phenotypic alterations pose significant risks of increased invasiveness, and thus will inform regarding the assignment of traits for which confinement will be necessary. This research should include reviews, analyses, and experimental tests of fundamental assumptions of ecological and evolutionary principles.

The committee briefly considered the ramifications of increased cooperation between public- and private-sector researchers. Intellectual property issues permeate current agricultural research and development, especially in biotechnology (NRC, 2003a), and research on bioconfinement methods is no exception. Indeed, development of bioconfinement methods to protect investment by preventing unlicensed use of GEOs has spurred much industry research in recent years (Chapter 3). Continued private support of applied research on some approaches to bioconfinement is to be expected, and the result should be the development of increasingly sophisticated and reliable methods.

Various changes during the past 20 years in U.S. law concerning intellectual property rights—together with political, economic, social, scientific, and technological developments—have led to increased collaboration between private industry and publicly funded research institutions. This has had favorable consequences, including bringing useful products to market more rapidly, raising new funds for public research and education, introducing academic scientists to the challenges of product development and the regulatory approval system, and providing access for academic researchers to proprietary information held by private industry. However, the increased mixing of public and private support has the potential to compromise fundamental agricultural research at public institutions—research which can be aimed uniquely at the public good (NRC, 1997; 2003a). The long-term effect could be to hamper the growth of the very research and innovation base upon which industry will rely (NRC, 1997). The committee emphasizes the need for scientists and administrators of publicly funded research programs to devise ways to work with industry for the public good, while at the same time recognizing their unique roles and importance in biological research.

References

Adams, J. M., G. Piovesan, S. Strauss, and S. Brown. 2002. The case for genetic engineering of native and landscape trees against introduced pests and diseases. Conservation Biology 16:874-879.

Adams, K. L., M. J. Clements, and J. C. Vaughn. 1998. The Peperomia cox I group I intron: Timing of horizontal transfer and subsequent evolution of the intron. Journal of Molecular Evolution 46:689-696.

Advani, S. J., R. R. Weichselbaum, R. J. Whitley, and B. Roizman. 2002. Friendly fire: Redirecting herpes simplex virus-1 for therapeutic applications. Clinical Microbiology and Infection 8(9):551-563.

Agricultural Biotechnology Research Advisory Committee (ABRAC). 1995. Working group on Aquatic Biotechnology and Environmental Safety for safely conducting research with genetically modified fish and shellfish. Parts I & II. USDA, Office of Agricultural Biotechnology, Documents No. 95 04-05. Washington, D.C. Available online at www.nbiap.vt.edu/perfstands/psmain.cfm. Accessed November 4, 2003.

Ahrenholtz, A., M. G. Lorenz, and W. Wackernagel. 1994. A conditional suicide system in *Escherichia coli* based on the intracellular degradation of DNA. Applied and Environmental Microbiology 60:3746-3751.

Al-Ahmad, H., and J. Gressel. 2002. Mitigation of transgene flow from crops to related weeds: Tobacco as a model. Pp. 229 in The 7th International Symposium on the Biosafety of Genetically Modified Organisms, Beijing, October 10-16.

Aleström, P., G. Kisen, H. Klungland, and O. Anderson. 1992. Fish gonadotropin-releasing hormone gene and molecular approaches for control of sexual maturation-development of a transgenic fish model. Molecular Marine Biology and Biotechnology 1(4-5):376-379.

Aleström, P. 1995. Genetic engineering in aquaculture. Pp. 125-130 in Proceedings of the First International Symposium on Sustainable Fish Farming, Oslo, Norway 28-31 August 1994. Reinertsen and Haaland, eds. The Netherlands: A.A.Balkema.

Allen, S. K., Jr., and X. Guo. 1998. The development and commercialization of tetraploid technology for oysters. Pp. 81-83 in Developments in Marine Biotechnology, Y. Le Gal and H. O. Halvorson, eds. New York: Plenum Press.

Allen, S. K., Jr., X. Guo, G. Burreson, and R. Mann. 1996. Heteroploid mosaics and reversion among triploid oysters, crassostrea gigas: Fact or artifact. Journal of Shellfish Research. 18:293.

Alpeter, F., J. Xu, and S. Ahmed. 2000. Generation of large numbers of independently transformed fertile perennial ryegrass (*Lolium prenne* L.) plants of forage- and turf-type cultivars. Molecular Breeding 6:519-528.

Alphey, L. 2002. Re-engineering the sterile insect technique. Insect Biochemical Molecular Biology 32(10):1243-1247.

Alphey, L., and M. Andreasen. 2002. Dominant lethality and insect population control. Molecular and Biochemical Parasitology 121(2):173-178.

Alvarez Morales, A. 2002. Transgenes in maize landraces in Oaxaca: Official report on the extent and implications. Pp. 78 in The 7th International Symposium on the Biosafety of Genetically Modified Organisms, Beijing, October 10-16.

Anderson, E. 1949. Introgressive Hybridization. New York: Wiley.

Apt, K. E., P. G. Kroth-Pancic, and A. Grossman. 1996. Stable nuclear transformation of the diatom *Phaeodactylum triconutum*. Molecular and General Genetics 252:572-579.

Aquagene, 2003. Company homepage. Online. Available at www.aquagene.com. Accessed February 23, 2003.

Arcand, F. 2003. Conference on plant-made pharmaceuticals, Quebec City, Quebec Canada, March 16-19. Available online at www.cpmp2003.org/pages/en/program/program.php. Accessed November 4, 2003.

Arnold, M. L. 1997. Natural Hybridization and Evolution. New York: Oxford University Press.

Arnold, M. L., and S. A. Hodges. 1995. Are natural hybrids fit or unfit relatives to their parents? Trends in Ecology and Evolution 10:67-71.

Arriola, P. E., and N. C. Ellstrand. 1997. Fitness of interspecific hybrids in the genus Sorghum: Persistence of crop genes in wild populations. Ecological Applications 7:512-518.

Asano, Y., and M. Ugaki. 1994. Transgenic plants of *Agrostis alba* obtained by electroporation-mediated direct gene transfer into protoplasts. Plant Cell Reports 13:243-246.

Atkinson, R. G., R. Schroder, I. C. Hallett, D. Cohen, and E. R. MacRae. 2002. Over-expression of polygalacturonase in transgenic apple trees leads to a range of novel phenotypes involving changes in cell adhesion. Plant Physiology 129:122-133.

Avri, A., and M. Edelman. 1991. Direct selection for paternal inheritance of chloroplasts in sexual crosses of Nicotiana. Molecular and General Genetics 225:273-277.

Baeumlein, H., A. Meuller, J. Schiemann, D. Helbing, R. Manteuffel, and U. Wobus. 1987. A lugumin B gene of *Vicia faba* is expressed in developing seeds of transgenic tobacco. Biologisches-Zantrablatt 10695:569-575.

Bahalla, P., I. Swoda, and M. Singh. 1999. Antisence-mediated silencing of a gene encoding a major ryegrass pollen allergen. Proceedings of the National Academy of Sciences of the USA 96:11676-11680.

Bakker, P. D., D. Glandorf, M. Viebahn, T. W. M. Ouivens, E. Smit, P. Leeflang, K. Wernars, L. S. Thomashow, J. E. Thomas-Oates, and L. C. van Loon. 2002. Effects of *Pseudomonas putida* modified to produce phenazine-1-carboxylic acid and 2,4-diacetylphloroglucinol on the microflora of field grown wheat. Antonie van Leuwenhoek 81:617-624.

Baroux C., R. Blanvillain, I. R. Moore, and P. Gallois. 2001. Transactivation of BARNASE under the AtLTP1 promoter affects the basal pole of the embryo and shoot development of the adult plant in Arabidopsis. Plant Journal 28(5):503-515.

Barrett, S. C. H. 1983. Crop mimicry in weeds. Economic Botany 37:255-282.

Bartley, D. M., L. Garibaldi, and R. L. Welcomme. 1998. Database on introductions of aquatic species. Available online at www.fao.org/fi/statist/fisoft/dias/index.htm. Accessed July 1, 2001.

Bartsch, D., U. Brand, C. Morak, M. Pohl, I. Schuphan, and N. C. Ellstrand. 2001. Biosafety aspects of genetically engineered virus resistant plants: performance of hybrids between transgenic sugar beet and Swiss chard. Ecological Applications 11:142-147.

Bateman, A. 1947. Contamination in seed crops. III. Relation with isolation distance. Heredity 1:303-306.

Baumberger, S., P. Dole, and C. Lapierre. 2002. Using transgenic poplars to elucidate the relationship between the structure and the thermal properties of lignins. Journal of Agricultural and Food Chemistry 50:2450-2453.

Beaumont, A. R., and J. E. Fairbrother. 1991. Ploidy manipulation in molluscan shellfish: A review. Journal of Shellfish Research 10:1-18.

Beclin, C., S. Boutet, P. Waterhouse, and H. Vaucheret. 2002. A branched pathway for transgene-induced RUA silencing in plants. Current Biology 12:684-688.

Bej, A., M. H. Perlin, and R. M. Atlas. 1988. Model suicide vector for containment of genetically engineered microorganisms. Applied Environmental Microbiology 54:2472-2477.

Benbrook, C. 2001. Do GM crops mean less pesticide use? Pesticide Outlook. Journal of The Royal Society of Chemistry 2001: 204-207.

Benfey, T. 1999. The physiology and behavior of triploid fishes. Reviews in Fisheries Science 7(1):39-67.

Benfey, T. J. 1989. A bibliography of triploid fish, 1943 to 1988. Report 1682. Winnipeg, Manitoba: Canada Data Report of Fisheries and Aquatic Sciences.

Berghammer, A. J., M. Klingler, and E. A. Wimmer. 1999. A universal marker for transgenic insects. Nature 402(6760):370-371.

Besnard, G., B. Khadari, P. Villemur, and A. Berville. 2000. Cytoplasmic male sterility in the olive (*Olea europaea* L.). Theoretical and Applied Genetics 100:1018-1024.

Beyer, P., S. Al-Babili, X. Ye, L. Paola, P. Schaub, R. Welsch, and I. Potrykus. 2002. Golden rice: Introducing the beta-carotene biosynthesis pathway into rice endosperm by genetic engineering to defeat vitamin A deficiency. Journal of Nutrition 132:506S-510S.

Bhatnagar, P., B. M. Glasheen, S. K. Bains, S. L. Long, R. Minocha, C. Walter, and S. C. Minocha. 2001. Transgenic manipulation of the metabolism of polyamines in poplar cells. Plant Physiology 125:2139-2153.

Bierzychudek, P. 1985. Patterns in plant parthenogenesis. Experientia 41:1255-1264.

Bishop-Hurley, S. L., R. J. Zabkiewicz, L. Grace, R. C. Gardner, A. Wagner, and C. Walter. 2001. Genetic transformation and hybridization: conifer genetic engineering: transgenic *Pinus radiata* (D. Don) and *Picea abies* (Karst) plants are resistant to the herbicide Buster. Plant Cell Reports 20:235-243.

Bjorkman, S. 1960. Studies in *Agrostis* and related genera. Symbolae Botanicae Upsalienses XVII 1:1-114.

Blonstein, A. D., T. Vahala, M. Koornneef, and P. J. King. 1988. Plants regenerated from auxin-auxotrophic variants are inviable. Molecular and General Genetics 215:58-64.

Bock, R. 2001. Transgenic plastids in basic research and plant biotechnology. Journal of Molecular Biology 312:425-438.

Bock, R. 2002. Modes of preventing gene flow due to outcrossing. Pp. 1-29 in The 7th International Symposium on the Biosafety of Genetically Modified Organisms, Beijing, October 10-16.

Boernke, F., M. Hajirezaei, and U. Sonnewald. 2002. Potato tubers as bioreactor for plalatinose production. Journal of Biotechnology 96(1):129-124.

Bolar, J. P., J. L. Norelli, G. E. Harman, S. K. Brown, and H. S. Aldwinckle. 2001. Synergistic activity of endochitinase from *Trichoderma atroviride (T. harzianum)* against the pathogenic fungus (*Venturia inaequalis)* in transgenic apple plants. Transgenic Research 10:533-543.

Boorse, Christopher. 1975. On the distinction between disease and illness. Philosophy and Public Affairs 5:49-68.

Borkovec, A. B. 1975. Control of insects by sexual sterilization. Environmental Letters 8(1):61-9.

Borkovec, A. B. 1976. Control and management of insect populations by chemosterilants. Environmental Health Perspectives 14(2):103-107.

Boudet, A. M., D. Goffner, C. Teulières, C. Marque, and J. Grima-Pettenati. 1998. Genetic manipulation of lignin profiles: A realistic challenge towards the qualitative improvement of plant biomass. AgBiotech News and Information 10:295-303.

Boudry, P., K. Broomberg, P. Saumitou-Laprade, M. Mörchen, J. Cuguen, and H. Van Dijk. 1994. Gene escape in transgenic sugar beet: What can be learned from molecular studies of weed beet populations? Pp. 75-87 in Proceedings of the Third International Symposium on the Bisafety Results of Field Tests of Genetically Modified Plants and Microorganisms, D. D. Jones, ed. Oakland: University of California Division of Agriculture and Natural Resources.

Boudry, P., M. Mörchen, P. Saumitou-Laprade, P. Vernet, and H. Van Dijk. 1993. The origin and evolution of weed beets: Consequences for the breeding and release of herbicide-resistant transgenic sugar beets. Theoretical and Applied Genetics 87:471-478.

Bouma, J. E., and R. E. Lenski. 1988. Evolution of a bacteria/plasmid association. Nature 335:351-352.

Braig, H. R., and G. Yan. 2002. The spread of genetic constructs in natural insect populations. Pp. 251-314 in Genetically Engineered Organisms: Assessing Environmental and Human Health Effects, D. K. Letourneau, and B. E. Burrows, eds. Boca Raton: CRC Press.

Bratspies, Rebecca. 2002. The illusion of care: Regulation, uncertainty, and genetically modified food crops. New York University School of Law Environmental Law Journal 10(3):297-355.

Briggs, H. 1999. GM pests bite back: Genetically modified food, the row continues. Science and Technology, BBC Science. Available online at http://news.bbc.co.uk/1/hi/sci/tech/337477.stm. Accessed November 4, 2003.

Brister, D. J. 2001. United States organic aquaculture: Moving towards national standards. World Aquaculture Society 32(3):51-53.

Broothaerts, W., P. A. Wiersma, and W. D. Lane. 2001. Genetics and genomics: multiplex PCR combining transgene and S-allele control primers to simultaneously confirm cultivar identity and transformation in apple. Plant Cell Reports 20:349-353.

Buchanan-Wollaston, V., J. E. Passiatore, and F. Cannon. 1987. The mob and oriT mobilization functions of a bacterial plasmid promote its transfer to plants. Nature 328:172-175.

Bull, J. J. 1994. Perspective: Virulence. Evolution 48(5):1423-1437.

Bundock, P., A. Den Dulk-Ras, A. Beijersbergen, and P. J. J. Hooykaas. 1995. Trans-kingdom T-DNA transfer from Agrobacterium tumefaciens to Saccharomyces cerevisiae. European Molecular Biology Organization Journal 14:3206-3214.

Burdon, J. J., D. R. Marshall, and N. H. Luig. 1981. Isozyme analysis indicates that a virulent cereal rust pathogen is a somatic hybrid. Nature 293:565-566.

Burgess, D. G., E. J. Ralston, W. G. Hanson, M. Heckert, M. Ho, T. Jenq, J. M. Palys, K. Tang, and N. Gutterson. 2002. A novel, two-component system for cell lethality and its use in engineering nuclear male-sterility in plants. The Plant Journal 31(1):113-125.

Burke, J. M., and L. H. Rieseberg. 2003. Fitness effects of transgenic disease resistance in sunflowers. Science 300:1250.

Burke, J. M., S. Tang, S. J. Knapp, and L. H. Rieseberg. 2002. Genetic analysis of sunflower domestication. Genetics 161:1257-1267.

Burnham, C. R. 1962. Discussions in Cytogenetics. St. Paul, Minn.: Charles R. Burnham.

Burns, D. R., R. Scarth, and P. B. E. Mcvetty. 1991. Temperature and genotypic effects on the expression of Po1 cytoplasmic male sterility in summer rape. Canadian Journal of Plant Science 71:655-661.

Burns, J. C., and T. T. Chen. 1999. Pantropic retroviral vectors for gene transfer in mollusks. U.S. Patent# 5,969,211.

Burns, J. C., and C. S. Friedman. 2002. Summary: Toward the genetic engineering of disease resistance in oysters. R/A-112. California: Sea Grant.

Burris, J. S. 2001. Adventitious pollen intrusion into hybrid maize seed production fields. Pp. 98-115 in Proceedings of the 56th Annual Corn and Sorghum Ind. Res. Conference, D. Wilkinson, ed. Washington, DC: American Seed Trade Association (ASTA).

Bushman, F. 2002. Lateral DNA Transfer Mechanisms and Consequences. New York: Cold Spring Harbor Laboratory Press.

Buxton, C. D., and P. A. Garrett. 1990. Alternative reproductive styles in seabreams (*Pisces Sparidae*). Environmental Biology of Fishes 28(1-4):113-124.

Cai, X., C. Brown, J. Adhiya, S. J. Traina, and R. T. Sayre. 1999. Growth and heavy metal binding properties of transgenic *Chlamydomonas* expressing a foreign metallothionein gene. International Journal of Phytoremediation 1:53-65.

Calvitti, M., M. Buttarazzi, C. Govoni, and U. Cirio. 1997. Use of induced sterility in adult *Trialeurodes vaporariorum* (Homoptera: Aleyrodidae) treated with gamma radiation. Journal of Economic Entomology 90(4):1022-1027.

Calvitti, M., P. C. Remotti, A. Pasquali, and U. Cirio. 1998. First results in the use of the sterile insect technique against *Trialeurodes vaporariorum* (Homoptera: Aleyroididae) in greenhouses. Annals of the Entomological Society of America 91(6):813-817.

Calvo, G. W., M. W. Luckenbach, S. K. Allen, Jr., and E. M. Burreson. 2001. Comparative field study of *Crassotrea gigas* and *Crassostrea virginica* in relation to salinity in Virginia. Journal of Shellfish Research 20:221-229.

Carlton, J. T. 1992. Dispersal of living organisms into aquatic ecosystems as mediated by aquaculture and fisheries activities. Pp. 13-46 in Dispersal of Living Organisms into Aquatic Ecosystems, A. Rosenfeld and R. Mann, eds. College Park: Maryland Sea Grant College.

Carpenter, J. E., K. A. Bloem, and S. Bloem. 2001. Applications of F1 sterility for research and management of *Cactoblastis cactorum* (Lepidoptera: Pyralidae). Florida Entomologist 84:531-536.

Carr, J. W., J. M. Anderson, F. G. Whoriskey, and T. Dilworth. 1997. The occurrence and spawning of cultured Atlantic salmon (*Salmo salar*) in a Canadian river. ICES (International Council for the Exploration of the Sea) Journal of Marine Science 54:1064-1073.

Catteruccia, F., T. Nolan, T. G. Loukeris, C. Blass, C. Savakis, F. C. Kafatos, and A. Crisanti. 2000. Stable germline transformation of the malaria mosquito *Anopheles stephensi*. Nature 405:959-962.

Cerutti, H., A. M. Johnson, N. W. Gillham, and J. E. Boynton. 1997. A eubacterial gene conferring spectinomycin resistance on *Chlamydomonas reinhardtii*: Integration into the nuclear genome and gene expression. Genetics 145:97-110.

Cervera, M., J. A. Pina, J. Juarez, L. Navarro, and L. Pena. 2000. A broad exploration of a transgenic population of citrus: Stability of gene expression and phenotype. Theoretical and Applied Genetics 100:670-677.

Chai, B., and M. Sticklen. 1998. Applications of biotechnology in turfgrass genetic improvement. Crop Science 38:1320-1338.

Chai, M., K. Senthil, S. Mo, Y. Chung, J. Shin, M. Park, and D. Kim. 2000. Embryogenic callus induction and Agrobacterium-mediated transformation of bentgrass (*Agrostis spp.*). Journal of Korean Society of Horticultural Science 41:450-454.

Chakrabarti, A., T. R. Ganapathi, P. K. Mukherjee, and B. A. Bapat. 2003. MSI-99, a magainin analogue, imparts enhanced disease resistance in transgenic tobacco and banana. Planta 216:587-596.

Charlesworth, B. 1989. The evolution of sex and recombination. Trends in Ecology and Evolution 4:264-267.

Chaudry, M. M., and J. M. Regenstein. 1994. Implications of biotechnology and genetic engineering for kosher and halal foods. Trends in Food Science and Technology 5:165-168.

Cheney, D. P., and C. Duke. 1995. Methods for producing improved strains of seaweed by fusion of spore-protoplasts, and resultant seaweeds and phycocolloids. U.S. Patent #5,426,040.

Cheng, Y. H., J. S. Yang, and S. D. Yeh. 1996. Efficient transformation of papaya by coat protein gene of papaya ringspot virus mediated by Agrobacterium following liquid-phase wounding of embryogenic ts with caborundum. Plant Cell Reports 16:127-132.

Chevassus, B. 1983. Hybridization in fish. Aquaculture 33:245-262.

Chilton, M. D., M. H. Drummond, D. J. Merlo, D. Sciaky, A. L. Montoya, M. P. Gordon, and E. W. Nester. 1977. Stable incorporation of plasmid DNA into higher plant cells: The molecular basis of crown gall tumorigenesis. Cell 11:263-271.

Cho, M., H. Choi, B. Buchanan, and P. Lemaux. 1999. Inheritance of tissue-specific expression of barley hordein promoter-uid fusions in transgenic barley plants. Theoretical and Applied Genetics 98(8):1253-1262.

Christou, P. 2002. No credible evidence is presented to support claims that transgenic DNA was introgressed into traditional maize landraces in Oaxaca, Mexico. Transgenic Research 11: iii-v.

Cipriani, G., R. Testolini, and M. Morgante. 1995. Paternal inheritance of plastids in interspecific hybrids of the genus Actinidia revealed by PCR-amplification of chloroplast DNA fragments. Molecular and General Genetics 247:693-697.

Clarke, D. K., E. A. Duarte, S. F. Elena, A. Moya, E. Domingo, and J. Holland. 1994. The red queen reigns in the kingdom of RNA viruses. Proceedings of the National Academy of Sciences of the USA 91(11):4821-4824.

Clifford, S. L., P. McGinnity, and A. Ferguson. 1998. Genetic changes in Atlantic salmon (Salmo salar) populations of Northwest Irish rivers resulting from escapes of adult farm salmon. Canadian Journal of Fisheries and Aquatic Sciences 55(2):358-363.

Coates, C. J., N. Jasinskiene, and A. A. James. 1998. Mariner transposition and transformation of the yellow fever mosquito, Aedes aegypti. Proceedings of the National Academy of Sciences of the USA 95(7):3748-3751.

Collares-Pereira, M. J. 1987. The evolutionary role of hybridization: The example of natural Iberian fish populations. Pp. 83-92 in K. Tiews, ed. Selection, Hybridization, and Genetic Engineering in Aquaculture. Berlin, Germany: Verlag H. Heenemann GmbH.

Colwell, R. E., E. A. Norse, D. Pimentel, F. E. Sharples, and D. Simberloff. 1985. Genetic engineering in agriculture. Science 229:111-112.

Committee, E. 2002. Standard Guide for Irradiation of Insects for Sterile Release Programs. West Conshohocken, PA: American Society for Testing Materials.

Connors, B. J., N. P. Laun, C. A. Maynard, and W. A. Powell. 2002. Molecular characterization of a gene encoding a cystatin expressed in the stems of American chestnut (Castanea dentata). Planta 215(3):510-514.

Contreras, A., S. Molin, and J. L. Ramos. 1991. Conditional-suicide containment system for bacteria which mineralize aromatics. Applied and Environmental Microbiology 57:1504-1508.

Cook, J. T., M. A. McNiven, G. F. Richardson, and A. M. Sutterlin. 2000. Growth rate, body composition and feed digestibility/conversion of growth-enhanced transgenic Atlantic salmon (Salmo salar). Aquaculture 188:15-32.

Corey, D. R., and J. M. Abrams. 2001. Morpholino antisense oligonucleotides: Tools for investigating vertebrate development. Genome Biology 2(5):1015.

Cotter, D., V. O. Donovan, N. O'Maioleidigh, G. Rogan, N. Roche, and N. P. Wilkins. 2000. An evaluation of the use of triploid Atlantic salmon (*Salmo salar* L.) in minimizing the impact of escaped farmed salmon on wild populations. Aquaculture 186:61-75.

Council on Environmental Quality and the Office of Science and Technology Policy. 2001. Case Studies of Environmental Regulation for Biotechnology. Available online at www.ostp.gov/html/012201.html. Accessed November 4, 2003.

Courtenay, W. R., Jr., and J. D. Williams. 1992. Dispersal of exotic species from aquaculture sources, with emphasis on freshwater fishes. Pp. 49-81 in Dispersal of Living Organisms into Aquatic Ecosystems, A. Rosenfeld and R. Mann, eds. College Park: Maryland Sea Grant College.

Crawley, M. J., and S. L. Brown. 1995. Seed limitation and the dynamics of oilseed rape on the M25 motorway. Proceedings of the Royal Society of London, Series B 259:49-54.

CSIRO (Commonwealth Scientific and Industrial Research Organization). 2002. Carp management in the Murray Darling Basin: Daughterless carp technology. Available online at www.marine.csiro.au/LeafletsFolder/pdfsheets/daughterless _carp_13may02.pdf. Accessed March 15, 2003.

Cummins, R., and M. Perlman, eds. 2002. Functions: New Essays in the Philosophy of Biology and Psychology. Oxford, UK: Oxford University Press.

Curtiss, R. III. 1988. Engineering organisms for safety: What is necessary? Pp. 7-20 in The Release of Genetically Engineered Micro-organisms, M. Sussman, G. H. Collins, F. A. Skinner, and D. E. Stewart-Tall, eds. London: Academic Press.

Curtiss, R. 1978. Biological containment and cloning vector transmissibility. Journal of Infectious Diseases 137:668-675.

Curtiss, III, R., M. Inoue, D. Pereira, J. C. Hsu, L. Alexander, and L. Rock. 1977. Construction and use of safer bacterial host strains for recombinant DNA research. Pp. 99-114 in Molecular Cloning of Recombinant DNA. New York: Academic Press.

Dale, P. J., B. Clarke, and E. M. G. Fontes. 2002. Potential for the environmental impact of transgenic crops. Nature Biotechnology 20:567-574.

Dalton, S., A. Bettany, E. Timms, and P. Morris. 1995. The effect of selection pressure on transformation frequency and copy number in transgenic plants of tall fescue (*Festuca arundinacea* Schreb.). Plant Science 108:63-70.

Daniell, H. 2002. Molecular strategies for gene containment in transgenic crops. Nature Biotechnology 20:581-586.

Daniell, H., and C. L. Parkinson. 2003. Jumping genes and containment. Nature 21:374-375.

Daniell, H., R. Datta, S. Varma, S. Gray, and S. B. Lee. 1998. Containment of herbicide resistance through genetic engineering of the chloroplast genome. Nature Biotechnology 16:345-348.

de Azevedo, J. F., R. Pinhao, and A. M. dos Santos. 1968. Biological studies carried out with the Glossina morsitans colony of Lisbon. Hypothetical control of the tsetse fly in Principe Island by the sterile-male technique. Anais da Escola Nacional de Saude Publicae de Medicina Tropical (Lisbon) 2(1):51-59.

De Buck, S., M. Van Montagu, and A. Depicker. 2001. Transgene silencing of invertedly repeated transgenes is released upon deletion of one of the transgenes involved. Plant Molecular Biology Reporter 46:433-445.

de Groot, M. J. A., P. Bundock, P. J. J. Hooykaas, and A. G. M. Beijersbergen. 1998. *Agrobacterium tumefaciens*-mediated transformation of filamentous fungi. Nature Biotechnology 16(9):839-842.

De Leij, F. A. A. M., E. J. Sutton, J. M. Whipps, J. S. Fenlon, and J. M. Lynch. 1995. Impact of field release of a genetically modified *Pseudomonas fluorescens* on indigenous microbial populations of wheat. Applied Environmental Microbiology 61:3443-3453.

De Pinto, M. C., F. Tommasi, and L. De Gara. 2002. Changes in the antioxidant systems as part of the signaling pathway responsible for the programmed cell death activated by nitric oxide and reactive oxygen species in tobacco Bright-Yellow 2 cells. Plant Physiology Preview 130:698-708.

DeFilippis, V. R., and L. P. Villarreal. 2000. An introduction to the evolutionary ecology of viruses. Pp. 125-208 in Viral Ecology, C. J. Hurst, ed. San Diego: Academic Press.

Devanas, M. A., and G. Stotzky. 1986. Fate in soil of a recombinant plasmid carrying a Drosophila gene. Current Microbiology 13:279-283.

Devlin, R. H., C. A. Biagi, T. Y. Yesaki, D. E. Smailus, and J. C. Byatt. 2001. Growth of domesticated transgenic fish. Nature 409:781-782.

DeWilde, C., N. Podevin, P. Windels, and A. Depicker. 2001. Silencing of antibody genes in plants with single-copy transgene inserts as a result of gene dosage effects. Molecular Genetics and Genomics 265:647-653.

Dominguez, A., C. Fagoaga, L. Navarro, P. Morena, and L. Pena. 2002. Regeneration of transgenic citrus plants under non selective conditions results in high-frequency recovery of plants with silenced genes. Molecular Genetics and Genomics 19:427-433.

Donaldson, E. M., and R. H. Devlin. 1996. Uses of Biotechnology to Enhance Production. Pp. 969-1020 in Principles of Salmonid Culture, W. Pennell and B. A. Barton, eds. Amsterdam: Elsevier Publishers.

Donaldson, E. M., R. H. Devlin, I. Solar, and F. Piferrer. 1993. The reproductive containment of genetically altered salmonids. Pp. 113-129 in Genetic Conservation of Salmonid Fishes, J. G. Cloud and G. H. Thorgaard, eds. New York: Plenum Press.

Dumas, S, and R. Campos Ramos. 1999. Triploidy induction in the Pacific white shrimp *Litopenaeus vannamei* (Boone). Aquaculture Research 30:621-624.

Dunahay, T. G. 1993. Transformation of *Chlamydomonas reinhardtii* with silicon carbide whiskers. Biotechniques 15:452-455.

Dunham, R. A., G. W. Warr, A. Nichols, P. L. Duncan, B. Argue, D. Middleton, and H. Kucuktas. 2002. Enhanced bacterial disease resistance of transgenic channel catfish *Ictalurus punctatus* possessing cecropin genes. Marine Biotechnology 4(3):338-344.

Durant, J., M. W. Bauer, and G. Gaskell, eds. 1998. Biotechnology in the Public Sphere: A European Sourcebook. London: Science Museum.

Eaton, D., F. van Tongeren, N. Louwaars, B. Visser, and I. Van der Meer. 2001. Economic and policy aspects of "terminator" technology. Biotechnology and Development Monitor 49:19-22.

Edlin, G., R. C. Tait, and R. L. Rodriguez. 1984. A bacteriophage lambda cohesive ends (COS) DNA fragment enhances the fitness of plasmid-containing bacteria growing in energy-limited chemostats. Biotechnology 2:251-254.

Edminster, C. June 5-9 2000. Future of turfgrass breeding techniques. Melbourne, Australia: Keystone address, at Australian Golf Course Superintendents Association Millennium Turfgrass Conference and Trade Exhibition.

Eichenwald, K., G. Kolata, and M. Peterson. 2001. Biotechnology food: From the lab to debacle. Originally in New York Times, 25 Jan. reprinted pp. 31-40 in Genetically Modified Foods: Debating Biotechnology, M. Ruse and D. Castle, eds. Amherst, New York: Prometheus Books.

El Euch, C., C. Jay-Allemand, M. Pastuglia, P. Doumas, J. P. Charpentier, P. Capelli, and L. Jouanin. 1998. Expression of antisense chalcone synthase RNA in transgenic hybrid walnut microcuttings: Effect of flavonoid content and rooting ability. Plant Molecular Biology 38:467-479.

El Malki, F., and M. Jacobs. 2001. Molecular characterization and expression study of a histidine auxotrophic mutant (his1) of *Nicotiana plumbaginifolia*. Plant Molecular Biology 45:191-199.

Elias, P. 2002. Regulators fine biotech companies for mishandling crops. Associated Press, Dec. 13.

Ellis, J. R., A. A. Shirsat, J. N. Yarwood, J. A. Gatehouse, R. R. D. Croy, and D. Boulter. 1988. Tissue-specific expression of a pea legume gene in seeds of *Nocotiana plumbaginifolia*. Plant Molecular Biology 10(3):203-214.

Ellstrand, N. C. 1988. Pollen as a vehicle for the escape of engineered genes? Pp. S30-S32 in Planned Release of Genetically Engineered Organisms, J. Hodgson and A. M. Sugden, eds. Cambridge, UK: Elsevier Publishers.

Ellstrand, N. C. 1989. Gene rustlers. Omni 11(7):33.

Ellstrand, N. C. 1992. Gene flow among seed plant populations. New Forests 6:241-256.

Ellstrand, N. C. 2003a. Dangerous liaisons? When Cultivated Plants Mate with Their Wild Relatives. Baltimore, MD: Johns Hopkins University Press.

Ellstrand, N. C. 2003b. Going to "great lengths" to prevent the escape of genes that produce specialty chemicals. Plant Physiology. In press.

Ellstrand, N. C., and C. A. Hoffman. 1990. Hybridization as an avenue for escape of engineered genes. BioScience 40:438-442.

Ellstrand, N. C., and D. R. Elam. 1993. Population genetic consequences of small population size: Implications for plant conservation. Annual Review Ecology Systematics 24:217-242.

Ellstrand, N. C., H. C. Prentice, and J. F. Hancock. 1999. Gene flow and introgression from domesticated plants into their wild relatives. Annual Review Ecology Systematics 30:539-563.

Ellstrand, N. C., R. Whitkus, and L. H. Rieseberg. 1996. Distribution of spontaneous plant hybrids. Proceedings of the National Academy of Sciences of the USA 93:5090-5093.

Epprecht, T. 1998. Genetic engineering and liability insurance. The Power of Public Perception. Zurich: SwissRe.

Eriksson, M. E., M. Israelsson, O. Olsson, and T. Moritz. 2000. Increased gibberellin biosynthesis in transgenic trees promotes growth, biomass production and xylem fiber length. Nature Biotechnology 18:784-788.

Ervin, D. E., R. Welsh, S. S. Batie, and C. L. Carpentier. 2003. Towards an ecological systems approach in public research for environmental regulation of transgenic crops. Review. Agriculture, Ecosystems and Environment 99:1-14.

Evans, M. M. S., and J. L. Kermicle. 2001. *Teosinte crossing barrier 1*, a locus governing hybridization of teosinte with maize. Theoretical and Applied Genetics 103:259-265.

Evenson, R. E., and D. Gollin. 2003. Assessing the impact of the Green Revolution, 1960 to 2000. Science 300:758-762.

Fabi, R. 2002. Update 2: Japan finds Starlink in US corn cargo-exporters. The Washington Post, December 27.

Fantes, J. A., and G. R. Mackay. 1978. The production of disomic addition lines of *Brassica campestris*. Cruciferae Newsletter 4:36-37.

FAO (Food and Agriculture Organization of the United Nations) Working Group on Plant Genetic Resources for Food and Agriculture. 2002. Available online at www.fao.org/ag/cgrfa/PGR.htm. Accessed November 4, 2003.

Fast, A. W., and P. Menasveta. 2000. Some recent issues and innovations in marine shrimp pond culture. Reviews in Fisheries Science 8(3):151-233.

Federal Register. 1986. Coordinated Framework for Regulation of the Products of Biotechnology. 51 Federal Register 23303. June 26.

Federal Register. 2003. Field testing of plants engineered to produce pharmaceutical and industrial compounds. Federal Register 68(46):11337-11340.

Fernandez-Cornejo, J., and W. D. McBride. 2002. Adoption of bioengineered crops. Agricultural Economic Report Number 810.Washington DC: USDA (U.S. Department of Agriculture), Economic Research Service.

Fillatti, J. J., J. Sellmer, B. Mccown, B. Haissig, and L. Comai. 1987. *Agrobacterium* mediated transformation and regeneration of *Populus*. Molecular and General Genetics 206:192-199.

Fire, A., S. Xu, M. K. Montgomery, S. A. Kostas, S. E. Driver, and C. C. Mello. 1998. Potent and specific genetic interference by double stranded RNA in *Caenorhabditis elegans*. Nature 391:806-811.

Fleming, I. A., K. Hindar, I .B. Mjølnerød, B. Jonsson, T. Balstad, and A. Lamberg. 2000. Lifetime success and interactions of farm salmon invading a native population. Proceedings of the Royal Society of London Biological Sciences 267(1452):1517-1523.

Fox, J. L. 2003. Puzzling industry response to Prodigene fiasco. Nature Biotechnology 21(1):3-4.

Fracheboud, Y., and P. J. King. 1988. Isolation of temperature-sensitive auxin-requiring variants of *Nicotiana plumbaginifolia*. Current Plant Science and Biotechnology in Agriculture 7:387-388.

Fracheboud, Y., and P. J. King. 1991. An auxin-auxotrophic mutant of *Nicotiana plumbaginifolia*. Molecular and General Genetics 227:397-400.

Fraga, M. F., R. Rodriguez, and M. J. Canal. 2002. Genomic DNA methylation-demethylation during aging and reinvigoration of *Pinus radiata*. Tree Physiology 22:813-816.

Fraser, P., M. Truesdale, C. Bird, W. Schuch, and P. Bramley. 1994. Carotenoid biosynthesis during tomato fruit development: Evidence of tissue-specific gene expression. Plant Physiology 105(1):405-413.

Fuhrman, J. A. 1999. Marine viruses and their biogeochemical and ecological effects. Nature 399:541-548.

Furner, I. J., G. A. Huffman, R. M. Amasino, D. J. Garfinkel, M. P. Gordon, and E. W. Nester. 1986. An *Agrobacterium* transformation in the evolution of the genus *Nicotiana*. Nature 319:422-427.

Futuyma, D. 1998. Evolutionary Biology. 3rd ed. Sunderland, MA: Sinauer Associates.

Galbreath, P. E., and B. L. Samples. 2000. Optimization of thermal shock protocols for induction of triploidy in brook trout. North American Journal of Aquaculture 62(4):249-259.

Garg, A. K., J. K. Kim, T. G. Owens, A. P. Ranwala, Y. D. Choi, L. V. Kochian, and R. J. Wu. 2002. Trehalose accumulation in rice plants confers high tolerance levels to different abiotic stresses. Proceedings of the National Academy of Sciences of the USA 99(25):15898-15903.

Gaskell, G., P. Thompson, and N. Allum. 2002. Worlds apart? Public opinion in Europe and the USA. Pp. 351-378 in Biotechnology: The Making of a Global Controversy, M. W. Bauer and G. Gaskell, eds. Cambridge, UK: Cambridge University Press.

Gebhard, F., and K. Smalla. 1999. Monitoring field releases of genetically modified sugar beets for persistence of transgenic plant DNA and horizontal gene transfer. Federation of European Microbiology Societies Microbiology Ecology 28:261-272.

Georghiou, G. P. 1986. The magnitude of the resistance problem. Pp. 14-43 in Pesticide Resistance: Strategies and Tactics for Management. Washington, DC: National Academy Press.

Gepts, P. 2001. Origins of plant agriculture and major crop plants. Pp. 629-637 in Our Fragile World: Challenges and Opportunities for Sustainable Development, M. K. Tolba, ed. Oxford, UK: Encyclopedia of Life Support Scientists Publishers.

Gerdes, K., P. B. Rasmussen, and S. Molin. 1986. Unique type of plasmid maintenance: Postsegregational killing of plasmid-free cells. Proceedings of the National Academy of Sciences of the USA 83:3116-3120.

Gervai, J., S. Peter, A. Nagy, L. Horvath, and V. Csanyi. 1980. Induction of triploidy in carp, *Cyprinus carpio* L. Journal of Fish Biology 17:667-671.

Ghorbel, R., R. Juarez, L Navarro, and L Pena. 1999. Green fluorescent protein as a screenable marker to increase the efficiency of generating transgenic woody fruit plants. Theoretical and Applied Genetics 99:350-358.

Ghosh, C., and P. L. Iversen. 2000. Intracellular delivery strategies for antisense phosphorodiamidate morpholino oligomers. Antisense Nucleic Acid Drug Development 10(4):263-274.

Giddings, G. 2000. Modeling the spread of pollen from *Lolium perenne*. The implications for the release of wind-pollinated transgenics. Theoretical and Applied Genetics 100:971-974.

Giddings, G., H. N. Sackville, and M. Hayward. 1997a. The release of genetically modified grasses. Part I: Pollen dispersal to traps in *Lolium perenne*. Theoretical and Applied Genetics 94:1000-1006.

Giddings, G., H. N. Sackville, and M. Hayward. 1997b. The release of genetically modified grasses. Part II: The influence of wind direction on pollen dispersal. Theoretical and Applied Genetics 94:1007-1014.

Gillis, J. 2002. EPA Fines Biotechs for Corn Violations. The Washington Post, Dec. 13 E3.

Ginder, R. 2001. Channeling, identity preservation and the value chain: Lessons from the recent problems with starlink corn. Available online at www.exnet.iastate.edu/Pages/grain/publications/starlink.html. Accessed November 4, 2003.

Gittins, J. R., T. K. Pellny, E. R. Hiles, C. Rosa, S. Biricolti, and D. J. James. 2000. Transgene expression driven by heterologous ribulose-1,5-bisphosphate carboxylase/oxygenase small-subunit gene promoters in the vegetative ts of apple (*Malus pumila* Mill.). Planta 210:232-240.

Glandorf, D. C. M., P. Verheggen, T. Jansen, J. W. Jorritsma, E. Smit, P. Leeflang, K. Wernars, L. S. Thomashow, E. Laureijs, J. E. Thomas-Oates, P. A. H. M. Bakker, and L. C. van Loon. 2001. Effect of genetically modified *Pseudomonas putida* WCS358r on the fungal rhizosphere microflora of field-grown wheat. Applied and Environmental Microbiology 67:3371-3378.

Goldbach, R. 1986. Molecular evolution of plant RNA viruses. Annual Review of Phytopathology 24:289-310.

Goodman, R. M., and N. Newell. 1985. Genetic engineering of plants for herbicide resistance: Status and prospects. Pp. 47-53 in Engineered Organisms in the Environment: Scientific Issues, H. O. Halvorson, D. Pramer, and M. Rogul, eds. Washington, DC: American Society of Microbiology.

Gossen, M., and H. Bujard. 1992. Tight control of gene expression in mammalian cells by tetracycline responsive promoters. Proceedings of the National Academy of the USA 89:5547-5551.

Gossen, M., S. Freundlieb, G. Bender, G. Muller, W. Hillen, and H. Bujard. 1995. Transcriptional activation by tetracycline in mammalian cells. Science 268:1766-1769.

Gracia, J. C., and J. R. Gonzalez. 1993. F-1 sterility of *Diatraea saccharalis* (Fab.), Lepidoptera: Crambidae: I. Effects of substerilizing doses on reproduction and competitiveness. Journal of Biochemistry 262:365-370.

Grant, V. 1981. Plant Speciation, 2nd ed. New York: Columbia University Press.

Gray, A., and A. Raybould. 1998. Reducing transgene escape routes. Nature 392:652-654.

Gray-Mitsumune, M., E. K. Molitor, D. Cukovic, J. E. Carlson, and C. J. Douglas. 1999. Developmentally regulated patterns of expression directed by poplar PAL promoters in transgenic tobacco and poplar. Plant Molecular Biology 39(4):657-669.

Green, M. B., H. M. LeBaron, and W. K. Moberg, eds. 1990. Managing Resistance to Agrochemicals: From Fundamental Research to Practical Strategies. American Chemical Society Symposium Series No. 421.

Gressel, J. 1999. Tandem constructs: Preventing the rise of superweeds. Trends in Biotechnology 17:361-366.

Gressel, J. 2001. Potential failsafe mechanisms against the spread and introgression of transgenic hypervirulent biocontrol fungi. Trends in Biotechnology 19:149-154.

Gressel, J. 2002. A proposed system for "Bio-barcoding"[TM] organisms.Beijing, China. Pp. 275 in Meeting Proceedings of the 7th International Symposium on the Biosafety of Genetically Modified Organisms.

Griffin, B. R. 1991. The U.S. Fish and Wildlife Service's triploid grass carp inspection program. Aquaculture Magazine Jan/Feb:188-189.

Gross, M. 1998. One species with two biologies: Atlantic salmon (*Salmo salar*) in the wild and in aquaculture. Canadian Journal of Fisheries and Aquatic Sciences 55(S1):131-144.

Gross, M. 2001. Potential impact of fish farming on wild salmon in British Columbia. Pp. 107-117 in The Swimmers: The Future of Wild Salmon on the North and Central Coasts of British Columbia. Victoria, British Columbia: Raincoast Conservation Society.

Grover, K. K., and H. V. Agarwal. 1980. Feasibility of using sterile-male technique for the control of *Anopheles stephensi* liston in the field: Laboratory and field cage studies with chemosterilants. Indian Journal of Experimental Biology 18(6):615-619.

Grunwald, C., F. Deutsch, D. Eckstein, and M. Fladung. 1999. Wood formation in rolC transgenic aspen trees. Trees 14:297-304.

Grunwald, C., K. Ruel, J. P. Joseleau, and M. Fladung. 2001. Morphology, wood structure, and cell wall composition of rolC transgenic and non-transformed aspen trees. Trees 15:503-517.

Guo, X. 1999. Superior growth as a general feature of triploid shellfish: Evidence and possible causes. Journal of Shellfish Research 18(1):266.

Guo, X. M., G. A. DeBrosse, and S. K. Allen. 1996. All-triploid pacific oyster (*Crassotrea gigas* Thunberg) produced by mating tetraploids and diploids. Aquaculture 142(3-4):149-161.

Ha, S., F. Wu, and T. Thorne. 1992. Transgenic turf-type tall fescue (*Festuca arundinacea* Schreb.) plants regenerated from protoplasts. Plant Cell Reports 11:601-604.

Hails, R. S. 2000. Genetically modified plants—the debate continues. Trends in Ecology and Evolution 15:14-18.

Hails, R., and J. Kinderlerer. 2003. The GM public debate: Context and communication strategies. Nature Reviews Genetics 4:819-824.

Hall, L., K. Topinka, J. Huffman, L. Davis, and A. Good. 2000. Pollen flow between herbicide-resistant *Brassica napus* is the cause of multiple-resistant *B. napus* volunteers. Weed Science 48:688-694.

Hallerman, E. M. 2003. Co-adaptation and outbreeding depression. Pp. 239-259 in Population Genetics: Principles and Applications for Fisheries Scientists, E. M. Hallerman, ed. Bethesda: American Fisheries Society.

Hallman, A., and M. Sumper. 1994. Reporter genes and highly regulated promoters as tools for transformation experiments in *Volvox carteri*. Proceedings of the National Academy of Sciences of the USA 91:11562-11566.

Halweil, Brian. 2000. Monsanto drops the terminator. Libraries Worldwide 739. World Watch 13:1:8-9.

Hammond, J., H. Lecoq, and B. Raccah. 1999. Epidemiological risks from mixed virus infections and transgenic plants expressing viral genes. Advances in Virus Research 54:189-314.

Hampp, R., M. Ecke, C. Schaeffer, T. Wallenda, A. Wingler, I. Kottke, and B. Sundberg. 1996. *Axenix mycorrhization* of wild type and transgenic hybrid aspen expressing T-DNA indoleacetic acid-biosynthetic genes. Trees 11:59-64.

Han, K. H., P. Fleming, K. Walker, M. Loper, W. S. Chilton, U. Mocek, M. P. Gordon, and H. G. Floss. 1994. Genetic transformation of mature *Taxus*: An approach to genetically control the in vitro production of the anticancer drug, taxol. Plant Science 95:187-196.

Hancock, J. F. 2003. A framework for assessing the risk of transgenic crops. BioScience 53:512-519.

Handel, S. N. 1983. Pollination ecology, plant population structure, and gene flow. Pp. 163-212 in Pollination Biology, L. Real, ed. New York: Academic Press.

Handler, A. M., and R. A. Harrell, II. 2001. Transformation of the Caribbean fruit fly, *Anastrepha suspensa*, with a piggyBac vector marked with polyubiquitin-regulated GFP. Insect Biochemistry and Molecular Biology 31(2):199-205.

Handler, A. M., S. D. McCombs, M. J. Fraser, and S. H. Saul. 1998. The lepidopteran transposon vector, piggyBac, mediates germ-line transformation in the Mediterranean fruit fly. Proceedings of the National Academy of Sciences of the USA 95(13):7520-7525.

Hanna, W. W. 1989. Characteristics and stability of a new cytoplasmic-nuclear male-sterile source in pearl millet. Crop Science 29:1457-1459.

Hansen, L. P., and B. Jonsson. 1991. The effect of timing of Atlantic salmon smolt and post-smolt release on the distribution of adult return. Aquaculture 98(1/3):61-67.

Hansen, S. R., and S. P. Hubbell. 1980. Single-nutrient microbial competition: Qualitative agreement between experimental and theoretically forecast outcomes. Science 207:1491-1493.

Hardee, D. D., and M. L. Laster. 1996. Current status of backcross sterility in *Heliothis virescens* (F.). Southwestern Entomologist 21(1):86-100.

Harl, N., R. Ginder, C. Hurlburgh, and S. Moline. 2001. The StarLink™ situation. Ames: Iowa State University. White Paper. Available online at www.exnet.iastate.edu/Pages/grain/publications/starlink.html. Accessed November 4, 2003.

Harrell, R. M., and W. Van Heukelem. 1998. A comparison of triploid induction validation techniques. The Progressive Fish Culturist 60:221-226.

Harries, H. C. 1995. Coconut. Pp. 57-62 in Evolution of Crop Plants, 2nd ed., J. Smartt and N. W. Simmonds, eds. Harlow, UK: Longman.

Hartl, D .L., and A. Clarke. 1997. Principles of Population Genetics, 3rd ed. Sunderland, MA: Sinaeur Association.

Hartl, D. L., D. E. Dykhuizen, R. D. Miller, L. Green, and J. DeFramond. 1983. Transposable element IS50 improves growth rate of *E. coli* cells without transposition. Cell 35:503-510.

Hartman, C., L. Lee, P. Day, and N. Tumer. 1994. Herbicide resistant turfgrass (*Agrostis palustris Huds.*) by Biolistic® transformation. Bio-Technology 12:919-923.

Hasan, M. M. 1999. Mating competitiveness of adult males of *Tribolium* spp (Coleoptera: Tenebrionidae) developing from irradiated pupae. Journal of Stored Products Research 35(3):307-316.

Haslberger, A. 2001. GMO contamination of seeds. Nature Biotechnology 19:613.

Hawkins, A. J. S., A. Gerard, M. Heral, and E. Zouros. 1998. Assessement of aquacultural advantages following the cytogenetic induction of polyploids in commercially important shellfish. Pp. 22-27 in K. G. Barthel, H. Barth, M. Bohle-Carbonell, C. Fragakis, E. Lipiatou, P. Martin, G. Ollier, and M. Weydert, eds. Third European Marine Science and Technology Conference (MAST Conference), Lisbon, 23-27 May: Project Synopses Vol. 5: Fisheries and Aquaculture (AIR: 1990-94) - Selected Projects from the Research Programme for Agriculture and Agro-Industry including Fisheries. Luxembourg: European Commission DG 12 Science, Research and Development.

Haygood, R., A. I. Ives, and D. A. Andow. 2003. Consequences of recurrent gene flow from crops to wild relatives. Proceedings of the Royal Society of London (B), Biological Sciences. In press.

Heasman, J. 2002. Morpholino oligos: Making sense of antisense? Developmental Biology 243(2):209-214.

Hebert, P. D. N., S. S. Schwartz, R. D. Ward, and T. L. Finston. 1993. Macrogeographic patterns of breeding system diversity in the Daphnia pulex group. 1. Breeding Systems of Canadian Populations. Heredity 70(2):148-161.

Hediger, M., M. Niessen, E. A. Wimmer, A. Dubendorfer, and D. Bopp. 2001. Genetic transformation of the housefly Musca domestica with the lepidopteran derived transposon piggyBac. Insect Molecular Biology 10(2):113-119.

Heifetz, P. 2000. Genetic engineering of chloroplast. Biochemisrty Paris 82: 655-666.

Heinemann, J. A., and G. F. Sprague, Jr. 1989. Bacterial conjugative plasmids mobilize DNA transfer between bacteria and yeast. Nature 340:205-209.

Heinrich, J. C., and M. J. Scott. 2000. A repressible female-specific lethal genetic system for making transgenic insect strains suitable for a sterile release program. Proceedings of the National Academy of Sciences of the USA 97:8229-8232.

Hershey, A. D., and R. Rotman. 1949. Genetic recombination between host range and plaque-type mutants of bacteriophage in single bacterial cells. Genetics 34:44-71.

Hew, C. L., and G. L. Fletcher. 1996. Transgenic salmonid fish expressing exogenous salmonid growth hormone. U.S. Patent #5,545,808.

Holefors, A., Z. T. Xue, L. H. Zhu, and M. Welander. 2000. Cell biology and morphogenesis: The Arabidopsis phytochrome B gene influences growth of the apple rootstock M26. Plant Cell Reports 19:1049-1056.

Hood, E. E., D. R. Witcher, S. Maddock, T. Meyer, C. Baszczynski, M. Bailey, P. Flynn, J. Register, L. Marshall, D. Bond, E. Kulisek, A. Kusnadi, R. Evangelista, Z. Nikolov, C. Wooge, R. Mehigh, R. Hernan, W. Kappel, D. Ritland, Li Chung Ping, and J. A. Howard. 1997. Commercial production of avidin from transgenic maize: Characterization of transformant, production, processing, extraction and purification. Molecular Breeding 3:291-306.

Horn, C., and E. A. Wimmer. 2003. A transgene-based, embryo-specific lethality system for insect pest management. Nature Biotechnology 21(1):64-70.

Horn, M., R. Shillito, B. Conger, and C. Harms. 1998. Transgenic plants of orchardgrass (Dactylis glomerata L.) from protoplasts. Plant Cell Reports 7:469-472.

Hu, G., and R. J. St. Leger. 2002. Field studies using a recombinant mycoinsecticide (Metarhizium anisopliae) reveal that it is rhizosphere competent. Applied and Environmental Microbiology 68:6383-6387.

Hu, W. J., S. A. Harding, J. Lung, J. L. Popko, J. Ralph, D. D. Stokke, C. J. Tsai, and V. L. Chiang. 1999. Repression of lignin biosynthesis promotes cellulose accumulation and growth in transgenic trees. Nature Biotechnology 17:808-812.

Huang, C. C., C. T. Lo, and K. H. Chang. 1974. Sex reversal in one sparid fish Chrysophrys major (Perciformes, Sparidae). Bulletin of the Institute of Zoology Academica Sinica 13:55-60.

Huang, C. V., M. A. Ayliffe, and J. N. Timmis. 2003. Direct measurement of the transfer rate of chloroplast DNA into the nucleus. Nature 422:72-76.

Hucl, P., and M. Matus-Cádiz. 2001. Isolation distances for minimizing out-crossing in spring wheat. Crop Science 41:1348-1351.

Huxel, G. R. 1999. Rapid displacement of native species by invasive species: Effect of hybridization. Biological Conservation 89:143-152.

Hwang Y., S. Nichol, S. Nandi, J. Jernstedt, and N. Huang. 2001. Aleuron and embryo-specific expression of the beta glucuronidase gene controlled by the barley Chi26 and Ltp1 promoters in transgenic rice. Plant Cell Reports 20(7):647-654.

Iida, A., A. Nagasawa, and K. Oeda. 1995. Positive and negative cis-regulatory regions in the soybean glycinin promoter identified by quantitative transient gene expression. Plant Cell Reports 14(9):539-544.

Ilardi, V., and M. Barba. 2001. Assessment of functional transgene flow in tomato fields. Molecular-Breeding 8:311-315.

Inada, Y., and N. Taniguchi. 1991. Spawning behavior and after-spawning survival in in-duced triploid ayu (in Japanese) Plecoglossus altivelis. Nippon Suisan Gakkaishi 57:2265-2269.

Inokuma, C., K. Sugiura, N. Imaizumi, and C. Cho. 1998. Transgenic Japanese lawngrass (Zoysia japonica Steud.) plants regenerated from protoplasts. Plant Cell Reports 17:334-338.

Inokuma, C., K. Sugiura, N. Imaizumi, C. Cho, and S. Kaneko. 1997. Transgenic Zoysia (Zoysia japonica) plants regenerated from protoplasts. International Turfgrass Society Research Journal 8:297-304.

Ito, J., A. Ghosh, L. A. Moreira, E. A. Wimmer, and M. Jacobs-Lorena. 2002. Transgenic anopheline mosquitoes impaired in transmission of a malaria parasite. Nature 417:452-455.

Jagannath, A., N. Arumugam, V. Gupta, A. Pradhan, P. K. Burma, D. Pental. 2002. Develop-ment of transgenic barstar lines and identification of a male sterile (barnase)/restorer (barstar) combination for heterosis breeding in Indian oilseen mustard (Brassica juncea). Current Science Bangalore 82(1):46-52.

Jambunathan, N., J. M. Siani, and T. W. McNellis. 2001. A humidity-sensitive Arabidopsis Copine mutant exhibits precocious cell death and increased disease resistance. Plant-Cell 13:2225-2240.

James, C. ed. 2002. Global review of commercialized transgenic crops: 2001. ISAAA Briefs: (26)1-184. Ithaca, NY: International Service for the Acquisition of Agri-biotech Applica-tions.

Jan, C. C. 2000. Cytoplasmic male sterility in two wild Helianthus annuus L. Accessions and their Fertility Restoration. Crop Science 40:1535-1538.

Jasinskiene, N., C. J. Coates, M. Q. Benedict, A. J. Comel., C. S. Rafferty, A. A. James, and F. H. Collins. 1998. Stable transformation of the yellow fever mosquito, Aedes aegypti, with the Hermes element from the housefly. Proceedings of the National Academy of Sciences of the USA 95(7):3743-3747.

Jefferson, R. A., T. A. Kavanagh, and M. W. Bevan. 1987. GUS fusions: ß-glucuronidase as a sensitive and versatile gene fusion marker in higher plants. European Molecular Biology Organisation Journal 6:3901-3907.

Jensen, L. B., J. L. Ramos, Z. Kaneva, and S. Molin. 1993. A substrate-dependent biological containment system for Pseudomonas putida based on the Escherichia coli gef gene. Applied and Environmental Microbiology 59:3713-3717.

Johnstone, R., R. M. Knott, A. G. Macdonald, and M. V. Walsingham. 1989. Triploidy induction in recently fertilized Atlantic salmon ova using anaesthetics. Aquaculture 78:229-236.

Jones, D A., M. H. Ryder, B. G. Clare, S. K. Farrand, and A. Kerr. 1988. Construction of a Tra- deletion mutant of pAgK84 to safeguard the biological control of crown gall. Molecular and General Genetics 212:207–214.

Kaplinsky, N., D. Braun, D. Lisch, A. Hay, S. Hake, and M. Freeling. 2002. Maize transgene results in Mexico are artifacts. Nature 416:601.

Kapuscinski, A. R. 2001. Controversies in designing useful ecological assessments of genetically engineered organisms. Pp. 385-415 in Genetically Engineered Organisms: Assessing Environmental and Human Health Effects, D. K. Letourneau and B. E. Burrows, eds. Boca Raton, FL: C.R.C Press.

Kapuscinski, A. R. 2003. Marine GEOs: Products in the pipeline. Marine Biotechnology Briefs 1(1):1-5, Table 1. Available online at www.fw.umn.edu/isees/MarineBrief/1/brief1.htm. Accessed November 4, 2003.

Kapuscinski, A. R., and D. J. Brister. 2001. Genetic impacts of aquaculture. Pp. 128-153 in Environmental Impacts of Aquaculture, K. D. Black, ed. Sheffield, UK: Sheffield Academic Press.

Kapuscinski, A. R., and E. M. Hallerman. 1991. Implications of introduction of transgenic fish into natural ecosystems. Canadian Journal of Fisheries and Aquatic Science 48:99-107.

Kapuscinski, A. R., and E. M. Hallerman. 1994. Benefits, environmental risks, social concerns, and policy implications of biotechnology in aquaculture. Contract report for Office of Technology Assessment, U.S. Congress. National Technical Information Service, U.S. Department of Commerce, PB96-197586.

Kapuscinski, A. R., R. M. Goodman, S. D. Hann, L. R. Jacobs, E. E. Pullins, C. S. Johnson, J. D. Kinsey, R. L. Kall, A. G. M. La Viña, M. Mellon, and V. W. Ruttan. 2003. Making 'safety first' a reality for biotechnology products. Nature Biotechnology 21(6):599-601.

Kargatova, T. V., E. E. Maksimova, and L. Yu. Popova. 2001. Coexistence of genetically engineered Escherichia coli strains and natural microorganisms in experimental aquatic microcosms. Microbiology 70(2):253-258.

Katsoyannos, B. I., N. T. Papadopoulos, N. A. Kouloussis, R. Heath, and J. Hendrichs. 1999. Method of assessing the fertility of wild Ceratitis capitata (Diptera: Tephritidae) females for use in sterile insect technique programs. Journal of Economic Entomology 92(3):590-597.

Kaufman, M. 2000. Biotech critics cite unapproved corn in taco shells; gene modified variety allowed only in animal feed because of allergy concerns. The Washington Post. Sept. 18, p. A02.

Keeler, K. H. 1989. Can genetically engineered crops become weeds? Biotechnology 7:1134-1139.

Keeler, K. H., and C. E. Turner. 1990. Management of transgenic plants in the environment. Pp. 189-218 in Risk Assessment in Genetic Engineering, M. Levin and H. Strauss, eds. New York: McGraw Hill.

Keenan, R. J., and W. P. C. Stemmer. 2002. Nontransgenic crops from transgenic plants. Nature Biotechnology 20:215-216.

Kelly, A. F., and R. A. T. George. 1998. Encyclopaedia of Seed Production of World Crops. Chichester, UK: John Wiley and Sons.

Kelso, D. T. 2003. The migration of salmon from nature to biotechnology. Chapter 3 in Engineering Trouble: Biotechnology and its Discontents, R. A. Schurmann and D. T. Kelso, eds. Berkeley: University of California Press.

Kenna, M. 2000. The United States Golf Association: A leader in the science of turf for 80 years. Diversity 16(1-2):11-14.

Kerremans, P., and G. Franz. 1995. Use of a temperature-sensitive lethal mutation strain of medfly (Ceratitis capitata) for the suppression of pest populations. Theoretical and Applied Genetics 90(3-4):511-518.

Keys, R. N., S. E. Smith, and H. L. Mogensen. 1995. Variation in generative cell plastid nucleoids and male fertility in Medicago sativa. Sexual Plant Reproduction 8:308-312.

Khadka, D. K., A. Nejidat, M. Tal, and A. Golan-Goldhirsh. 2002. DNA markers for sex: Molecular evidence for gender dimorphism in dioecious *Mercurialis annua* L. Molecular breeding: New Strategies in Plant Improvement 9:251-257.

Khan, M. S., and P. Maliga. 1999. Fluorescent antibiotic resistance marker for tracking plastid transformation in higher plants. Nature Biotechnology 17:910-915.

Kistner, A., M. Gossen, F. Zimmermann, J. Jerecic, C. Ullmer, H. Lubbert, and H. Bujard. 1996. Doxycycline-mediated quantitative and tissue-specific control of gene expression in transgenic mice. Proceedings of the National Academy of Sciences of the USA 93:10933-10938.

Kitamura, S., H. Ogata, and H. Onozato, 1991. Triploid male masu salmon *Oncorhynchus masou* shows normal courtship behavior. Nippon Suisan Gakkaishi 57:2157.

Kitchell, J. F., and S. W. Hewett. 1987. Forecasting forage demand and yields of sterile Chinook salmon *(Oncorhynchus tshawytscha)* in Lake Michigan. Canadian Journal of Fisheries and Aquatic Sciences 44(Suppl. 5):284-290.

Klein, T. M., E. C. Harper, Z. Svab, J. C. Sanford, M. E. Fromm, and P. Maliga. 1988. Stable genetic transformation of intact *nicotiana* cells by the particle bombardment process. Proceedings of the National Academy of Sciences of the USA 85:8502-8505.

Knipling, E. F. 1968. The potential role of sterility for pest control. Principles of Insect Chemosterilization, G. C. Labrecque and C. N. Smith, eds. New York: Appleton-Century-Crofts.

Knudsen, S. M., and O. H. Karlstrom. 1991. Development of efficient suicide mechanisms for biological containment of bacteria. Applied and Environmental Microbiology 57:85-92.

Knudsen, S., P. Saadbye, L. H. Hansen, A. Collier, B. L. Jacobsen, J. Schlundt, and O. H. Karlstrom. 1995. Development and testing of improved suicide functions for biological containment of bacteria. Applied and Environmental Microbiology 61:985-991.

Kolpin, D. W., E. T. Furlong, M. T. Meyer, E. M. Thurman, S. D. Zaugg, L. B. Barber, and H. T. Buxton. 2002. Pharmaceuticals, hormones, and other organic wastewater contaminants in U.S. Streams, 1999-2000: A national reconnaissance. Environmental Science and Technology 36(6):1202-1211.

Koonin, E. V., K. S. Makarova, and L. Aravind. 2001. Horizontal gene transfer in prokaryotes: Quantification and classification. Annual Review of Microbiology 55:709-742.

Krafsur, E. S. 1998. Sterile insect technique for suppressing and eradicating insect population:55 years and counting. Journal of Agricultural Entomology 15(4):303-317.

Kramer, M. G., and K. Redenbaugh. 1994. Commercialization of a tomato with an antisense polygalacturonase gene – The Flavr Savr^{TM} tomato story. Euphytica 79:293-297.

Kreuzer, H., and A. Massey. 2001. Recombinant DNA and Biotechnology: A Guide for Teachers. 2nd ed. Washington, DC: ASM Press.

Krohne, D. T., I. Baker, and H. G. Baker. 1980. The maintenance of the gynodioecious breeding system in Plantago-Lanceolata. American-Midland-Naturalist 103(2):269-279.

Kumar, S., and M. Fladung. 2001. Gene stability in transgenic aspen *(Populus)*. II. Molecular characterization of variable expression of transgene in wild and hybrid aspen. Planta 13:731-740.

Kumari, S. L., and M. Mahadevappa. 1998. Genetic, histological and histochemical evidence for reversion to partial fertility in WA CMS line IR54752A. International Rice Research Notes 23(2):9.

Kunik, T., T. Tzfira, Y. Kapulnik, Y. Gafni, C. Dingwall, and V. Citovsky. 2001. Genetic transformation of HeLa cells by *Agrobacterium*. Proceedings of The National Academy of Sciences of the USA 98:1871–1876.

Kusnadi, A. R., Z. L. Nikolov, and J. A. Howard. 1997. Production of recombinant proteins in transgenic plants. Biotechnology Bioengineering 56:473-484.

Kuvshinov, V., K. Koivu, A. Kanerva, and E. Pehu. 2001. Molecular control of transgene escape from genetically modified plants. Plant Science 160:517-522.

Lagler, K. F., J. E. Bardach, and R. R. Miller. 1977. Ichthyology. New York: John Wiley and Sons.

Lambrecht, B. 2001. Dinner at the New Gene Café. New York: St. Martin's Press.

Lee, L. 1996. Turfgrass biotechnology. Plant Science 115:1-8.

Lee, L., C. Laramore, P. Day, and N. Tumer. 1996. Transformation and regeneration of creeping bentgrass (*Agrostis palusteris Huds.*) protoplasts. Crop Science 36:401-406.

Leffel, S., S. A. Mabon, and C. N. Stewart, Jr. 1997. Tracking transgenic plants using green fluorescent protein. Pp. 378-380 in Biotechnology Risk Assessment: 8~ Symposium of the Environmental Risks of Biotechnology Products: Risk Assessment Methods and Research Progress, M. Levin, C. Grim, and J. S. Angle, eds. Ottawa, Canada.

Lefol, E., A. Fleury, and H. Darmency. 1996. Gene dispersal from transgenic crops. II. Hybridization between oilseed rape and the wild hoary mustard. Sexual Plant Reproduction 9:189-196.

Lenski, R.E., and T. T. Nguyen. 1988. Stability of recombinant DNA and its effects on fitness 6:518-520. Trends in Biotechnology/Trends in Ecology and Evolution Special Publication. Cambridge, UK: Elsevier Publications.

Lever, C. 1996. Naturalized Fishes of the World. San Diego: Academic Press.

Levin, B. R., and R. E. Lenski. 1983. Coevolution in bacteria and their viruses and plasmids, Pp. 99-127 in Coevolution, D. J. Futuyma and M. Slatkin, eds. Sunderland, MA: Sinauer Associates.

Levin, D. A. 1978. Genetic variation in annual phlox self compatible vs. self incompatible species. Evolution 32:245-263.

Levin, D. A. 2002. The Role of Chromosomal Change in Plant Evolution. Oxford, UK: Oxford University Press.

Levin, D. A. 2003. The congener as an agent of extermination and rescue of rare species. Evolutionary Conservation Biology, R. Ferriere, U. Dieckmann, and D. Covet, eds. Cambridge, UK: Cambridge University Press.

Levin, D. A., and H. W. Kerster. 1975. The effect of gene dispersal on the dynamics and statics of gene substitution in plants. Heredity 35:317-336.

Levin, D. A., J. Francisco-Ortega, and R. K. Jansen. 1996. Hybridization and the extinction of rare plant species. Conservation Biology 10:10-16.

Levine, I., K. Watson, and D. Cheney. 2001. Marine agronomy – The introduction of exotic or modified cultivars: is the risk worth the reward? Pp. 132-139 in Marine Aquaculture and the Environment, A Meeting for Stakeholders in the Northeast, M. Tlusty, D. Bengston, H. O. Halvorson, S. Oktay, J. Pearce, and R. B. Rheault, Jr., eds. Falmouth, MA: Cape Cod Printing.

Lev-Yadun, S., and R. Sederoff. 2001. Grafting for transgene containment. Nature Biotechnology 19:1104.

Li, W.L., J. D. Faris, S. Muthukrishnan, D. J. Liu, P. D. Chen, and B. S. Gill. 2001. Isolation and characterization of novel cDNA clones of acidic chitinases and beta-1,3-glucanases from wheat spikes infected by Fusarium graminearum. Theoretical and Applied Genetics 102(2/3):353-362.

Li, F., J. Xiang, L. Zhou, C. Wu, and X. Zhang. 2003. Optimization of triploid induction by heat shock in Chinese shrimp *Fenneropenaeus chinensis*. Aquaculture 219:221-231.

Liang, L. N., J. L. Sinclair, L. M. Mallory, and M. Alexander. 1982. Fate in model ecosystems of microbial species of potential use in genetic engineering. Applied and Environmental Microbiology 44:708-714.

Lin, W., G. K. Price, and E. Allen. 2001. StarLink: Impacts on the U.S. corn market and world trade. Feed Yearbook/April 2001. Economic Research Service, USDA: 40-48. Available online at www.ers.usda.gov/Briefing/Biotechnology/starlinkarticle.pdf. Accessed November 4, 2003.

Lin, Y. 2001. Risk assessment of bar gene transfer from B and D genomes of transformed wheat (Triticum aestivum) lines to jointed goatgrass (Aegilops cylindrica). Journal of Anhui Agricultural University 28(2).

Linares, B. C. 1985. Male sterility in *Alnus glutinosa*. Silvae-Genetica 34:69-72.

Linder, C. R., I. Taha, G. J. Seiler, A. A. Snow, and L. H. Rieseberg. 1998. Long-term introgression of crop genes into wild sunflower populations. Theoretical and Applied Genetics 96:339-347.

Liu, C. A. 1996. Genetic engineering of creeping bentgrass for resistance to the herbicide glufosinate ammonia. Ph.D. Dissertation. East Lansing: Michigan State University.

Liu, S., Y. Liu, G. Zhu, X. Zhang, C. Luo, H. Feng, X. He. G. Zhu, and H. Yang. 2001. The formation of tetraploid stocks of red crucian carp x common carp hybrids as an effect of interspecific hybridization. Aquaculture 192:171-186.

Lloyd, D. G. 2000. The selection of social actions in families: III. Reproductively disabled individuals and organs. Evolutionary Ecology Research 2:29-40.

Longden, P. C. 1993. Weed beet: a review. Aspects of Applied Biology 35:185-194.

Lord, E. M. 1981. Cleistogamy: A tool for the study of floral morphogenesis, function, and evolution. The Botanical Review 47:421-449.

Losey, J. E., L. S. Rayor, and M. E. Carter. 1999. Transgenic pollen harms monarch larvae. Nature 3999:214.

Losoff, R. M., H. L. Mogensen, T. Zhu, and S. E. Smith. 1995. The zygote and prembryo of alfalfa: Quantitative, three-dimensional analysis and implications for biparental plastid inheritance. Protoplasma 189:88-100.

Loukeris, T. G., B. Arca, I. Livadaras, G. Dialektaki, and C. Savakis. 1995. Introduction of the transposable element Minos into the germ line of *Drosophila melanogaster*. Proceedings of the National Academy of Sciences of the USA 92(21):9485-9489.

Lowe-McConnell, R. H. 2000. The Roles of Tilapias in Ecosystems. Pp. 129-162 in Tilapias: Biology and Exploitation, M. C. M. Beveridge and B. J. Andrews, eds. Dordrecht, Netherlands: Kluwer Academic Press.

Lu, B. 2003. Transgene containment by molecular means: Is it possible and cost effective? Environmental Biosafety Research. In press.

Lu, Bao-Rong., Z. P. Song, and J. K Chen. 2003. Can transgenic rice cause ecological risks through transgene escape? Progress in Natural Science 13:17-24.

Lu, J. K., T. T. Chen, S. K. Allen, T. Matsubara, and J. C. Burns. 1996. Production of transgenic dwarf surfclams, *Mulinia lateralis*, with pantropic retroviral vectors. Proceedings of National Academy of Sciences of the USA 93:3482-3486.

Lura, H., and H. Seagrov. 1991. Documentation of successful spawning of escaped farmed female Atlantic salmon, *Salmo salar*, in Norwegian rivers. Aquaculture 98(1/3):151-159.

Lux, S. A., and K. Gaggl. 1996. Ethological analysis of medfly courtship: Potential for quality control. Pp. 425-432, In Economic Fruit Flies: A World Assessment of Their Biology and Management, B. A. McPheron and G. J. Steck, eds. Delray Beach, FL: St. Lucia Press.

Mack, R. N., and M. Eisenberg. 2002. The United States naturalized flora: Largely the product of deliberate introductions. Annals of the Missouri Botanical Garden 89:176-189.

MacKay, J. J., D. M. O'Malley, T. Presnell, F. L. Booker, M. M. Campbell, R. W. Whetten, and R. R. Sederoff. 1997. Inheritance, gene expression, and lignin characterization in a mutant pine deficient in cinnamyl alcohol dehydrogenase. Proceedings of the National Academy of Sciences of the USA 94:8255-8260.

Maclean, N., and R. J. Laight. 2000. Transgenic fish: An evaluation of benefits and risks. Fish and Fisheries 1:146-172.

Madigan, M. T., J. M. Martinko, and J. Parker. 2003. Brock Biology of Microorganisms. Upper Saddle River, NJ: Prentice Hall.

Magnus, D., and A. Caplan. 2002. The primacy of the moral in the GMO debate. Pp. 80-87 in Genetically Modified Foods: Debating Biotechnology, Castle and Ruse, eds. Amherst, NY: Prometheus Press.

Magnus, D., A. Caplan, and G. McGee, eds. 2002. Who Own Life? Amherst, NY: Prometheus Press.

Makee, H., and G. Saour 1999. Nonrecovery of fertility in partially sterile male *Phthorimaea operculella* (Lepidoptera: Gelechiidae). Journal of Economic Entomology 92(3):516-520.

Maliga, P. 2001. Plastid engineering bears fruit. Nature Biotechnology 19:826-827.

Maliga, P. 2002. Engineering the plastic genome of higher plants. Current Opinion in Plant Biology 5:164-172.

Maliga, P. 2003. Progress towards commercialization of plastid transformation technology. Trends in Biotechnology 21(1):20-28.

Mann, C. C. 2002. Has GM corn "invaded" Mexico? Science 295:1617-1619.

Mann, C. C., and M. L. Plummer. 2002. Forest Biotech Edges Out of the Lab. Science 295:1626-1629.

Mansour, Y. M., and E. S. Krafsur. 1991. Induction of dominant lethal mutations for control of the facefly, *Musca autumnalis* DeGeer. Medical and Veterinary Entomology 5(2):175-81.

Marcus, J. P., K. C. Goulter, J. L. Green, S. J. Harrison, and J. M. Manners. 1997. Purification, characterisation and cDNA cloning of an antimicrobial peptide from *Macadamia inegrifolia*. European Journal of Biochemistry 15:743-749.

Marshall, B., P. Flynn, D. Kamely, and S. B. Levy. 1988. Survival of *Escherichia coli* with and without ColE1:Tn5 after aerosol dispersal in a laboratory and a farm environment. Applied and Environmental Microbiology 54:1776-1783.

Marshall, D. R., and A. H. D. Brown. 1981. The evolution of apomixis. Heredity 47:1-15.

Marsula, R., and C. Wissel. 1994. Insect pest control by a spatial barrier. Ecological Modeling 75-76:203-211.

Martinez, R., J. Juncal, C. Zaldivar, A. Arenal, I. Guillen, V. Morera, O. Carrillo, M. Estrada, A. Morales, and M. P. Estrada. 2000. Growth efficiency in transgenic tilapia (*Oreochromis sp.*) carrying a single copy of a homologous cDNA growth hormone. Biochemical and Biophysical Research Communications 267:466-472.

Martinez-Soriano, J. P. R., A. M. Bailey, J. Lara-Reyna, and D. S. Leal-Klevezas. 2002. Transgenes in mexican maize. Nature Biotechnology 20:19.

Marvier, M. 2001. Ecology of Transgenic Crops. American Scientist 89(2):160-167.

Marvier, M. A., E. Meir, and P. M. Kareiva. 1999. How do the design of monitoring and control strategies affect the chance of detecting and containing transgenic weeds? Pp. 109-122 in Risks and Prospects of Transgenic Plants, Where Do We Go From Here? K. Ammann and Y. Jacot, eds. Basel, Switzerland: Birkhauser Press.

Mayell, H. 2002. Maryland Wages War on Invasive Walking Fish. National Geographic News, July 2. Available online at http://news.nationalgeographic.com/news/2002/07/0702_020702_snakehead.html. Accessed November 4, 2003.

Mayfield, S. 2003. Expression of human antibodies in eukaryotic algae. Conference on Plant-Made Pharmaceuticals, Quebec City, Canada, March 16-19. Available online at www.cpmp2003.org/pages/en/program/program.php. Accessed April 28, 2003.

Maynard Smith, J. 1978. The evolution of sex. Cambridge, UK: Cambridge University Press.

Mazithulela, G., D. Sudhakar, T. Heckel, L. Mehlo, P. Christau, J. Davis, and M. Boulton. 2000. The maize streak virus coat protein transcription unit exhibits tissue-specific expression in transgenic rice. Plant Science 155(1):21-29.

McCabe, M. S., U. B. Mohapatra, S. C. Debnath, J. B. Power, and M. R. Davey. 1999. Integration, expression and inheritance of two linked T DNA marker genes in transgenic lettuce. Molecular Breeding: New Strategies in Plant Improvement 5:329-344.

McClure, R. 2002. State's ban on gene-altered fish a first. Seattle Post-Intelligencer Reporter, Monday December 23. Available online at http://seattlepi.nwsource.com/local/ 101109fish23.shtml. Accessed December 27, 2002.

McGinnity, P., C. Stone, and A. Ferguson. 1997. Genetic impact of escaped farmed Atlantic salmon (*Salmo salar L.*) on native populations: Use of DNA profiling to assess freshwater performance of wild, farmed, and hybrid progeny. ICES (International Council for the Exploration of the Sea) Journal of Marine Science 4(6):998-1008.

McHughen, A., G. G. Rowland, F. A. Holm, R. S. Bhatty, E. O. Kenaschuk. 1997. CDC Triffid transgenic flax. Canadian Journal of Plant Science 77:641-643.

McHughen, Alan. 2000. Pandora's Picnic Basket. Oxford, UK: Oxford University Press.

McInnis, D. O., S. Tam, C. Grace, and D. Miyashita. 1994. Population suppression and sterility rates induced by variable sex ratio, sterile insect releases of *Ceratitis capitata* (Diptera: Tephritidae) in Hawaii. Annals of the Entomological Society of America 87(2):231-240.

McIntyre, G. R. 2002. The application of system safety engineering and management techniques at the US Federal Aviation Adminiistration (FAA). Safety Science 40:325-335.

McLean, M. A., and P. J. Charest. 2000. The regulation of transgenic trees in North America. Silvae Genetica 49:233-239.

McQuillan, A. G. 2000. Cabbages and Kings: The ethics and aesthetics of new forestry. Pp. 293-318 in Environmental Ethics and Forestry: A Reader, Peter List, ed. Philadelphia, PA: Temple University Press.

McQuillan, A. G. 2001. Naturalness as a value: How forest biotechnology might be affected by public views on forests and science. Pew Initiative on Biotechnology and Food conference on Forestry and Biotechnology. Available online at http://pewagbiotech.org/events/ 1204/presentations/McQuillan2.pdf. Accessed November 4, 2003.

Meilan, R., D. J. Auerbach, C. Ma, S. P. DiFazio, and S. H. Strauss. 2002. Stability of herbicide resistance and GUS expression in transgenic hybrid poplars (*Populus* sp.) during several years of field trials and vegetative propagation. HortScience 37(2):277-280.

Meilan, R., K. H. Han, C. Ma, S. P. DiFazio, J. A. Eaton, E. A. Hoien, B. J. Stanton, R. P. Crockett, M. L. Taylor, R. R. James, J. S. Skinner, L. Jouanin, G. Pilate, and S. H. Strauss. 2002. The CP4 transgene provides high levels of tolerance to Roundup® herbicide in field-grown hybrid poplars. Canadian Journal of Forest Research 32:967-976.

Meinke, D. W. 1991. Embryonic mutants of *Arabidopsis thaliana*. Developmental Genetics 12:382-392.

Merkle, S. A., and J. F. Dean. 2000. Forest tree biotechnology. Current Opinion in Biotechnology 11:298-302.

Mett, V., E. Podivinski, A. Tennant, L. Lochhead, W. Jons, and P. Reynolds. 1996. A system for tissue-specific copper-controllable gene expression in transgenic plants: Nodule-specific antisense of aspartate aminotransferase P2. Transgenic Research 5(2):105-113.

Metz, M., and J. Fütterer. 2002. Suspect evidence of transgenic contamination. Nature 416: 600-601.

Metz, P. L. J., E. Jacobsen, J. P. Nap, A. Pereira, and W. J. Stiekema. 1997. The impact of biosafety of the phosphinothricin-tolerance transgene in inter-specific B. *rapa* times B. *napus* hybrids and their successive backcrosses. Theoretical and Applied Genetics 95:442-450.

Metzlaff, M. 2002. RNA-Mediated RNA degradation in transgene-and virus-induced gene silencing. Biological Chemistry 383(10):1483-1489.

Meza, T. J., B. Stangeland, I. S. Mercy, M. Skarn, D. A. Nymoen, A. Berg, M. A. Butenko, A. M. Hakelien, C. Haslekas, L. A. Meza-Zepeda, and R. B. Aalen. 2002. Analyses of single-copy *Arabidopsis* T-DNA-transformed lines show that the presence of vector backbone sequences, short inverted repeats and DNA methylation is not sufficient or necessary for the induction of transgene silencing. Nucleic Acids Research 30:4556-4566.

Michalik, B. 1978. Stability of the male sterility in carrots under different growth conditions. Bulletin. Serie Des Sciences Biologiques 26:827-832.

Milks, M. L., M. Leptich, and D. A. Theilmann. 2001. Recombinant and wild-type nucleopolyhedroviruses are equally fit in mixed infections. Environmental Entomology 30:972-981.

Miller, L. M., A. R. Kapuscinski, and W. Senanan. In Press. A biosafety approach to addressing risks posed by aquaculture escapees. In Biosafety and Environmental Impact of Genetic Enhancement and Introduction of Improved Strains of Tilapia / Exotics in Africa. Penang, Malaysia: International Center for Living Aquatic Resources Management Conference Proceedings of the International Center for Living Aquatic Resources Management.

Minocha, S. C. 2003. Genetic engineering of seaweeds: current status and perspectives. Pp. 19-26 in The Seventeenth International Seaweed Symposium, Proceedings of the 18th International Seaweed Symposium, A. R.O. Chapman, R. J. Anderson, V. J. Vreeland, and I. R. Davison, eds. Cape Town, South Africa, January 28 – February 2. Oxford, UK: Oxford University Press.

Mittler, R., and L. Rizhsky. 2000. Transgene-induced lesion mimic. Plant Molecular Biology 44:335-344.

Mogensen, H., and M. Rusche. 2000. Occurrence of plastids in rye sperm cells. American Journal of Botany 87:1189-1192.

Mohanty A. K., U. K. Mukhopadhyay, S. Grover, and V. K. Batish. 1999. Bovine chymosin: Production by rDNA technology and application in cheese manufacture. Biotechnology Advances 17(2-3):205-217.

Molin, S. P., P. Klemm, L. K. Poulsen, H. Biehl, K. Gerdes, and P. Andersson. 1987. Conditional suicide system for containment of bacteria and plasmids. Bio/Technology 5:315-1318.

Molin, S., L. Boe, L. B. Jensen, C. S. Kristensen, M. Givskov, J. L. Ramos, and A. K. Bej. 1993. Suicidal genetic elements and their use in biological containment of bacteria. Annual Review of Microbiology 47:139-166.

Monsanto Canada, Inc., and Monsanto Co. v. Schmeiser and Schmeiser, LTD. 2002. FCA 309. On appeal to the Canadian Supreme Court as of January 2004.

Moore, W. S. 1984. Evolutionary Ecology of Unisexual Fishes. Pp. 329-398 in Evolutionary Genetics of Fishes. B.J. Turner, ed. New York: Plenum Press.

Moss, S. M., G. S. Pluem, B. J. Argue, S. M. Arce, D. R. Moss, and D. J. Lopiccolo. 2003. Effects of exogenous teroid hormones on the sex ratio and survival of Pacific white shrimp Litopenaeus vannamei. Aquaculture America Book of Abstracts. Louisville, Kentucky: World Aquaculture Society.

Muir, W. M., and R. D. Howard. 1999. Possible ecological risks of transgenic organism release when transgenes affect mating success: Sexual selection and the Trojan gene hypothesis. Proceedings of the National Academy of Sciences of the USA 92:13853-13856.

Muir, W. M., and R. D. Howard. 2001. Fitness components and ecological risk of transgenic release: a model using Japanese medaka (Oryzias latipes). The American Naturalist 158:1-16.

Muir, W. M., and R. D. Howard. 2002. Assessment of possible ecological risks and hazards of transgenic fish with implications for other sexually reproducing organisms. Transgenic Research 11:101-114.

Mundt, C. C. 1995. Models from plant pathology on the movement and fate of new genotypes of microorganisms in the environment. Annual Review of Phytopathology 33:467-488.

Murphy, B. R., and R. M. Chanock. 2001. Immunization against viral diseases. Pp. 435-467 in Fields Virology, P. M. Howley, ed. Philadelphia: Lippincott Williams and Wilkins.

Nam, Y. K., H. J. Cho, Y. S. Cho, J. K. Noh, C.G. Kim, and D. S. Kim. 2001a. Accelerated growth, gigantism and likely sterility in autotransgenic triploid mud loach *Misgurnus mizolepis*. Journal of the World Aquaculture Society 32(4):353-363.

Nam, Y. K., J. K. Noh, Y. S. Cho, H. J. Cho, K. N. Cho, C. G. Kim, and D. S. Kim. 2001b. Dramatically accelerated growth and extraordinary gigantism of transgenic mud loach *Misgurnus mizolepis*. Transgenic Research 10:353-362.

National Commission for the Protection of Human Subjects of Biomedical and Behavioral Research. 1979. The Belmont Report: Ethical Principles and Guidelines for the Protection of Human Subjects Research. Available online at *http://ohrp.osophs.dhhs.gov/humansubjects/guidance/belmont.htm*. Accessed November 4, 2003.

NRC (National Research Council). 1983. Risk Assessment in the Federal Government: Managing the Process. Washington, DC: National Academy Press.

NRC (National Research Council). 1987. Introduction of Recombinant DNA-Engineered Organisms into the Environment: Key Issues. Washington, DC: National Academy Press.

NRC (National Research Council). 1989a. Field Testing Genetically Modified Organisms: Framework for Decisions. Washington, DC: National Academy Press.

NRC (National Research Council). 1989b. Investing in Research: A Proposal to Strengthen the Agricultural, Food, and Environmental System. Washington, DC: National Academy Press.

NRC (National Research Council). 1996. Understanding Risk: Informing Decisions in a Democratic Society. Washington, DC: National Academy Press.

NRC (National Research Council). 1997. Intellectual Property Rights and Plant Biotechnology: Proceedings of a Forum. Washington, DC: National Academy Press.

NRC (National Research Council). 2000. Genetically Modified Pest-Protected Plants: Science and Regulation. Washington, DC: National Academy Press.

NRC (National Research Council). 2002a. Environmental Effects of Transgenic Plants: The Scope and Adequacy of Regulation. Washington, DC: National Academy Press.

NRC (National Research Council). 2002b. Animal Biotechnology: Science-Based Concerns. Washington, DC: The National Academies Press.

NRC (National Research Council). 2002c. Publicly Funded Agricultural Research and the Changing Structure of U. S. Agriculture. Washington, DC: National Academy Press.

NRC (National Research Council). 2002d. Genetic Status of Atlantic Salmon in Maine. Interim Report from the Committee on Atlantic Salmon in Maine. Washington, DC: National Academy Press.

NRC (National Research Council). 2003a. Frontiers in Agricultural Research: Food, Health, Environment, and Communities. Washington, DC: The National Academies Press.

NRC (National Research Council). 2003b. Countering Agricultural Bioterrorism. Washington, DC: The National Academies Press.

NRC (National Research Council). 2004. Atlantic Salmon in Maine. Washington, DC: The National Academies Press.

Naylor, R. L., S. L. Williams, and D. R. Strong. 2001. Aquaculture: A gateway for exotic species. Science 294:1655-1656.

Nell, J. 2002. Farming triploid oysters. Aquaculture 210:69-88.

Nestle, M. 2003. Safe Food. Berkeley, CA: University of California Press.

Nielsen, K. M., A. M. Bones, K. Smalla, and J. D. van Elsas. 1998. Horizontal gene transfer from transgenic plants to terrestrial bacteria – A rare event? Fedaration of Eurpoean Microbiology Societies Microbiology Reviews 22:79-103.

Nielsen, K. M., J. D. van Elsas, and K. Smalla. 2000. Transformation of *Acinetobacter* sp. Strain BD413 (pFG4D*nptII*) with transgenic plant DNA in soil microcosms and effects of kanamycin on selection of transformants. Applied and Environmental Microbiology 66:1237-1242.

No, E., Y. Zhou, and C.A. Loopstra. 2000. Sequences upstream and downstream of two xylem-specific pine genes influence their expression. Plant Science 160:77-86.

Nowak, R. 2002. Gene warfare to be waged on invasive fish. New Scientist, May 8. Available online at www.newscientist.com/news. Accessed May 15, 2002.

Occhipinti-Ambrogi A., D. Savini. 2003. Biological invasions as a component of global change in stressed marine ecosystems. Marine Pollution Bulletin 46(5):542-551.

O'Connell, A., K. Holt, J. Piquemal, J. Grima-Pettenati, A. Boudet, B. Pollet, C. Lapierre, M. Petit-Conlil, W. Schuch, and C. Halpin. 2002. Improved paper pulp from plants with suppressed cinnamoyl-CoA reductase or cinnamyl alcohol dehydrogenase. Transgenic Research 11:495-503.

O'Toole, G., H. B. Kaplan, and R. Kolter. 2000. Biofilm formation as microbial development. Annual Review of Microbiology 54:49-79.

Office of Science and Technology Policy and Council on Environmental Quality. 2001. Case Studies of Environmental Regulation for Biotechnology: Case study number one: Growth-enhanced salmon. Available online at www.ostp.gov/html/ceq_ostp_study2.pdf. Accessed November 4, 2003.

Ohkawa, Y. 1984. Cytoplasmic male sterility in *Brassica-Campestris*-Ssp-*Rapifera*. Japanese Journal of Breeding 34:285-294.

PABE (Public Perceptions of Agricultural Biotechnology in Europe). 2001. Final report of the PABE research project 88.

Palese, P. 1984. Reassortment continuum, Pp. 141-151 in Concepts in Viral Pathogenesis. A. L. Notkins and M. B. A. Oldstone, eds. New York: Springer-Verlag.

Palmer, J. D., K. L. Adams, Y. Cho, C. L. Parkinson, Y.-L. Qiu, and K. Song. 2000. Dynamic evolution of plant mitochondrial genomes: Mobile genes and introns and highly variable mutation rates. Pp. 35-57 in Variation and Evolution in Plants and Microorganisms: Toward a New Synthesis 50 Years After Stebbins, F. J. Ayala, M. W. Fitch, and M. T. Clegg, eds. Washington D.C: National Academy Press.

Palmer, S., V. S. Scanferlato, D. R. Orvos, G. H. Lacy, and J. Cairns. 1997. Survival and ecological effects of genetically engineered *Erwinia carotovora* in soil and aquatic microcosms. Environmental Toxicology and Chemistry 16 (4):650-657.

Park, Y., and H. Cheong. 2002. Expression and production of recombinant human interleukin-2 in potato plants. Protein Expression and Purification. 25(1):160-165.

Parker, I. M., D. Simberloff, W. M. Lonsdale, K. Goodell, M. Wonham, P. M. Kareiva, M. H. Williamson, B.Von Holle, P. B. Moyle, J. E. Byers, and L. Goldwasser. 1999. Impact: Toward a framework for understanding the ecological effects of invaders. Biological Invasions 1:3-19.

Pedersen, J. F., D. B. Marx, and D. L. Funnell. 2003. Use of A3 cytoplasm to reduce risk of gene flow through sorghum pollen. Crop Science 43(4):1506-1509.

Peloquin, J. J., and T. A. Miller. 2000. Enhanced green fluorescent protein DNA in pink bollworm (*Pectinophora gossypiella*) through polymerase chain reaction amplification. Journal of Cotton Science 4:28-33.

Peloquin, J. J., S. T. Thibault, R. T. Staten, and T. A. Miller. 2000. Germ-line transformation of pink bollworm (Lepidoptera: Gelechiidae) mediated by the *piggyBac* transposable element. Insect Molecular Biology 9(3):323-333.

Pena, L., and A. Seguin. 2001. Recent advances in the genetic transformation of trees. Trends in Biotechnology 19:500-506.

Pena, L., M. Martin-Trillo, J. Juarez, J. A. Pina, L. Navarro, and J. M. Martinez-Zapater. 2001. Constitutive expression of Arabidopsis LEAFY or APETALA1 genes in citrus reduces their generation time. Nature Biotechnology 19:263-267.

Pesticide and Toxic Chemical News. 2003. StarLink lawsuit settlement seen as warning to industry. 31(17).

Pew Initiative on Food and Biotechnology. 2002. Pharming the field: A look at the benefits and risks of bioengineering plants to produce pharmaceuticals. Washington, DC: Pew Initiative on Food and Biotechnology. Proceedings of a workshop sponsored by the Pew Initiative on Food and Biotechnology, the US FDA (Food and Drug Administration), and the CSREES (Cooperative State Research, Education, and Extension Service) of the U.S. Department of Agriculture.

Pew Initiative on Food and Biotechnology. 2003. Pp. 72 in Future Fish: Issues in Science and Regulation of Transgenic Fish. Washington, DC: Pew Initiative on Food and Biotechnology

Piers, K. L., J. D. Heath, K. M. Liang, K. M. Stephens, and E.W. Nester. 1996. *Agrobacterium tumefaciens*-mediated transformation of yeast. Proceedings of the National Academy of Sciences of the USA 93:1613-1618.

Pieters, A. J. 1897. Seed production and seed saving. Pp. 207-216 in Yearbook of the USDA, 1896. Washington, DC: Government Printing Office. 1895-1920.

Pitkänen, T. I., A. Krasnov, M. Reinisalo, and H. Mölsä. 1999. Transfer and expression of glucose transporter and hexokinase genes in salmonid fish. Aquaculture 173:319-332.

Poirier, Y., D. Dennis, K. Klomparens, and C. Somerville. 1992. Polyhydroxybutyrate, a biodegradable thermoplastic produced in transgenic plants. Science 256:520-423.

Pollack, A. 2001. Altered corn surfaced earlier. The New York Times, Sept. 4.

Pontier, D., S. Gan, R. M. Amasino, D. Roby, and E. Lam. 1999. Markers for hypersensitive response and senescence show distinct patterns of expression. Plant Molecular Biology 9:1243-1255.

Poppendieck, H. H., and J. Petersen. 1999. A longtime experiment in dispersal biology: The naturalization of the mistletoe (*Viscum album L.*) in Hamburg and adjacent regions. Abhandlungen-Naturwissenschaftlichen-Verein-zu-Bremen 44:377-396.

Potrykus I., M. Saul, J. Petruska, J. Paszkowski, and R. Shillito. 1985. Direct gene transfer to cells of a graminaceous monocot. Molecular and General Genetics 199:183-188.

Pratt, L. A., and R. Kolter. 1998 .Genetic analysis of *Escherichia coli* biofilm formation: Roles of flagella, motility, chemotaxis and type I pili. Molecular Microbiology 30(2):285-293.

Preston, N. P., V. J. Baule, R. Leopold, J. Henderling, P. W. Atkinson, and S. Whyard. 2000. Delivery of DNA to early embryos of the Kuruma prawn, *Penaeus japonicus*. Aquaculture 181:225-234.

Priest, S. 2001. A grain of truth. New York: Rowman and Littlefield.

Puterka, G. J., C. Bocchetti, P. Dang, R. L. Bell, and R. Scorza. 2002. Pear transformed with a lytic peptide gene for disease control affects nontarget organism, pear psylla (Homoptera: Psyllidae). Journal of Economic Entomology 95:797-802.

Qaim, M., and D. Zilberman. 2003. Yield effects of genetically modified crops in developing countries. Science 299:900-902.

Quist, D., and I. H. Chapela. 2001. Transgenic DNA introgressed into traditional maize landraces in Oaxaca, Mexico. Nature 414:541-543.

Quist, D., and I. H. Chapela. 2002. Quist and Chapela reply. Nature 416: 602.

Qureshi, Z. A., T. Hussain, and N. Ahmed. 1993. Evaluation of the F-1 sterility technique for population suppression of the pink bollworm, *Pectinophora gossypiella* (Saunders) in Management of Insect Pests: Nuclear and Related Molecular Genetic Techniques, P. Haward-Kitto, R. F. Kelleher, and G. V. Ramesh, eds. Vienna, Austria: International Atomic Energy Agency.

Rahman, M. A., A. Ronyai, B. Z. Engidaw, K. Jauncey, G. L. Hwang, A. Smith, E. Roderick, D. Penman, L. Varadi, and N. Maclean. 2001. Growth and nutritional trials on transgenic Nile tilapia containing an exogenous fish growth hormone gene. Journal of Fish Biology 59:62-78.

Ralph, J., C. Lapierre, F. Lu, J. M. Marita, G. Pilate, W. Van Doorsselaere, J. Boerjan, and L. Jouanin. 2001. NMR evidence for benzodioxane structures resulting from incorporation of 5-hydroxyconiferyl alcohol into Lignins of O-methyltransferase-deficient poplars. Journal of Agricultrual and Food Chemistry 49:3508.

Rambaut, A., D. Posada, K. A. Crandall, and E. C. Holmes. 2004. The causes and consequences of HIV evolution. Nature Reviews Genetics 5:52-61.

Rapoport, H. F., and L. Rallo. 1990. Ovule development in normal and parthenocarpic olive fruits. Acta Horticulturae 286:223-226.

Raufner, M. 2001. Designer trees. Biotechnology and Development Monitor 44/45:2-7.

Ravelonandro, M., R. Scorza, and J. Dunez. 1998. Characterization of phenotype resistance to plum pox of transgenic plants expressing plum pox virus capsid gene. Acta Virologica 42:70-272.

Razak, S. A., G. L Hwang, M. A. Rahman, and N. Maclean. 1999. Growth performance and gonadal development of growth enhanced transgenic tilapia Oreochromis niloticus (L.) following heat-shocked-induced triploidy. Marine Biotechnology 1:533-544.

Reamon-Buttner, S. M., J. Schondelmaier, and C. Jung.1998. AFLP markers tightly linked to the sex locus in Asparagus officinalis L. Molecular breeding: New strategies in plant improvement 4:91-98.

Reganold, J. P., J. D. Glover, P. K. Andrews, and H. R. Hinman. 2001. Sustainability of three apple production systems. Nature 410:926-930.

Renn, D. 1997. Biotechnology and the red seaweed polysaccharide industry: Status, needs and prospects. Trends in Biotechnology 15:9-14.

Reznick, D. N., F. H. Shaw, F. H. Rodd, and R. G. Shaw. 1997. Evaluation of the rate of evolution in natural populations of guppies (Poecilia reticulata). Science 275:1934-1937.

Rhymer, J. M., and D. Simberloff. 1996. Extinction by hybridization and introgression. Annual Review of Ecology and Systematics 27:83-109.

Richards, A. J. 1997. Plant breeding systems. 2nd ed. London: Chapman and Hall.

Rissler J., and M. Mellon. 1996. The Ecological Risks of Engineered Crops. Cambridge, MA: Massachusetts Institute of Technology Press.

Robinson, A. S. 2002. Mutations and their use in insect control. Mutatation Research 511(2):113-132.

Robleto, E. A., J. Borneman, and E. W. Triplett. 1998. Effects of bacterial antibiotic production on rhizosphere microbial communities from a culture: Independent perspective. Applied and Environmental Microbiology 64:5020-5022.

Roland. H. E., and B. Moriarity. 1990. System Safety Engineering and Management. New York: John Wiley.

Rong, Y. S. and K. G. Golic. 2000. Gene targeting by homologous recombination in Drosophila. Science 288(5473):2013-2018.

Rong, Y. S., and K. G. Golic 2001. A targeted gene knockout in Drosophila. Genetics 157(3):1307-1312.

Rong, Y. S., S. W. Titen, H. B. Xie, M. M. Golic, M. Bastiani, P. Bandyopadhyay, B. M. Olivera, M. Brodsky, G. M. Rubin, and K. G. Golic. 2002. Targeted mutagenesis by homologous recombination in D. melanogaster. Genes and Development 16(12):1568-81.

Rosati, C., A. Cadic, M. Duron, M. Ingouff, and P. Simoneau. 1999. Molecular characterization of the anthocyanin synthase gene in Forsythia intermedia reveals organ-specific expression during flower development. Plant Science 149(1):73-79.

Rose, N. F., P. A. Marx, A. Luckay, D. F. Nixon, W. J. Moretto, S. M. Donahoe, D. Montefiori, A. Roberts, L. Buonocore, and J. K. Rose. 2001. An effective AIDS vaccine based on live attenuated vesicular stomatitis virus recombinants. Cell 106:539-549.

Rosenfeld, A., and R. Mann. 1992. Pp. 471 in Dispersal of Living Organisms into Aquatic Ecosystems. College Park: Maryland Sea Grant College.

Rosewich, V. L., and H. C. Kistler. 2000. Role of horizontal gene transfer in the evolution of fungi. Annual Review of Phytopathology 38:325-363.

Ruf, S., M. Hermann, I. J. Berger, H. Carrer, and R. Bock. 2001. Stable genetic transformation of tomato plastids – High-level foreign protein expression in fruits. Nature Biotechnology 19:870-875.

Rusche, M., H. Mogensen, T. Zhu, and T. Smith. 1995. The zygote and prembryo of alfalfa. Protoplasma 189:88-100.

Sakuta, C., and S. Satoh. 2000. Vascular tissue-specific gene expression of xylem sap glycine-rich proteins in roots and their localization in the walls of metaxylem vessels in cucumber. Plant and Cell Physiology 41(5):627-638.

Salyers, A. A., and D. D. Whitt. 2002. Bacterial Pathogenesis: A Molecular Approach, 2nd ed. Washington, DC: American Society for Microbiology Press.

Saner, M. 2001. Real and metaphorical moral limits in the biotech debate. Nature Biotechnology 19:609.

Sarmasik, A., I. K. Jang, C. Z. Chun, J. K. Lu, and T. T. Chen. 2001. Transgenic live-bearing fish and crustaceans produced by transforming immature gonads with replication-defective pantropic retroviral vectors. Marine Biotechnology 3(5):470-477.

Scanferlato, U. S., D. R. Orvos, J. Cairns, and G. H. Lacy. 1989. Genetically engineered Erwinia carotovora in aquatic microcosms: survival and effects on functional groups of indigenous bacteria. Applied and Environmental Microbiology 55:1477-1482.

Scheffler, J. A., R. Parkinson, and P. J. Dale. 1993. Frequency and distance of pollen dispersal from transgenic oilseed rape (Brassica napus). Transgenic Research 2:356-364.

Schernthaner, J. P., S. F. Fabijanski, P. G. Arnison, M. Racicot, and L. S. Robert. 2003. Control of seed germination in transgenic plants based on the segregation of a two-component genetic system. Proceedings of the National Academy of Sciences of the USA 100(11):6855-6859.

Schiedlmeier, B., R. Schmitt, W. Muller, M. Kirk, H. Gruber, W. Mages, and D. Kirk. 1994. Nuclear transformation of Volvox carteri. Proceedings of the National Academy of Sciences of the USA 91:5080-5084.

Schultheis, J. R., and S. Walters. 1998. Yield and virus resistance of summer squash cultivars and breeding lines in North Carolina. HortTechnology 8:31-39.

Schülter, K., J. Fütterer, and I. Potrykus. 1995. Horizontal, gene transfer from a transgenic potato line to a bacterial pathogen (Erwinia chrysanthemi) occurs, if at all, at an extremely low frequency. Biotechnology 13:1094-1098.

Schumann, C., and J. Hancock. 1991. Paternal inheritance of plastids in Medicago sativa. Theoretical and Applied Genetics 78:863-866

Scientists' Working Group on Biosafety. 1998. Manual for Assessing Ecological and Human Health Effects of Genetically Engineered Organisms. Part One: Introductory Text and Supporting Text for Flowcharts. Part Two: Flowcharts and Worksheets. Edmonds, WA: The Edmonds Institute. Available online at www.edmonds-institute.org/manual.html. Accessed November 4, 2003.

Screen, S. E., and R. J. St. Leger. 2000. Cloning expression and substrate specificity of a fungal chymotrypsin. Journal of Biological Chemistry 275:6689-6694.

Sears, M. K., R. L. Hellmich, D. R. Stanley-Horn, K. S. Oberhauser, J. M. Pleasants, H. R. Mattila, B. D. Siegfried, and G. P. Dively. 2001. Impact of Bt corn pollen on monarch butterfly populations: A risk assessment. Proceedings of the National Academy of Sciences of the USA 98:11937-11942.

Sederoff, R. 1999. Building better trees with antisense. Nature Biotechnology 17:750-751.

Seth, P. K., and S. S. Sehgal 1993. Partial sterilizing radiation dose effect on the F-1 progeny of *Spodoptera litura* (Fabr.): Growth, bioenergetics and reproductive competence. Pp. 427-440 in Proceedings, Management of Insect Pests: Nuclear and Related Molecular Genetics Techniques, P. Howard-Kitto, R. F. Kelleher, and G. V. Ramesh, eds. Vienna, Austria: International Atomic Energy Agency.

Shi, Y., and F. V. Hebard. 1997. Male sterility in progeny derived from hybridizations between *Castanea dentata* and *C. mollissima*. Journal of the American Chestnut Foundation 11:38-47.

Sidorov, V., D. Kasten, S. Pang, P. Hajdukiewicz, J. Staub, and N. Nehra. 1999. Stable chloroplast transformation in potato: Use of green fluorescent protein as a plasmid marker. Plant Journal 19(2):209-216.

Siegele, D. A., and R. Kolter. 1992. Life After Log. Journal of Bacteriology 174:345-348.

Sikdar S., G. Srino, S. Chaudhuri, and P. Maliga. 1998. Plastid transformation in *Arabidopsis thaliana*. Plant Cell Reports 18:20-24.

Simberloff, D. 2003. Eradication: Preventing invasions at the outset. Weed Science 51:247-253.

Simmonds, N. W. 1995. Bananas. Pp. 370-374 in Evolution of Crop Plants, 2nd ed. J. Smartt and N. W. Simmonds, eds. Harlow, UK: Longman.

Singer, M., and D. Soll. 1973. Guidelines for DNA Hybrid Molecules. Science 181:1114.

Skinner, M. W., and B. M. Pavlik. 1994. Inventory of rare and endangered vascular plants of California. Sacramento, CA: California Native Plant Society.

Slatkin, M. 1987. Gene flow and the geographic structure of natural populations. Science 236:787-792.

Slavov, G. T., S. P. DiFazio, and S. H. Strauss. 2002. Gene flow in forest trees: From empirical estimates to transgenic risk assessment. Pp. 113-133 in Gene Flow Workshop, The Ohio State University, March 5 and 6.

Smith, N. A., S. P. Singh, M. B. Wang, P. Stoutjesdijk, A. Green, and P. M. Waterhouse. 2000. Total silencing by intron-spliced hairpin RNAs. Nature 407:319-320.

Snow, A., A. D. Pilson, L. H. Rieseberg, M. J. Paulsen, N. Pleskac, M. R. Reagon, D. E. Wolf, and S. M. Selbo. 2003. A *Bt* transgene reduces herbivory and enhances fecundity in wild sunflowers. Ecological Applications 13(2):279-286.

Snow, A., and P. Moran-Palma. 1997. Commercialization of transgenic plants: Potential ecological risks. BioScience 47:86-96.

Sobecky, P. A., M. A. Schell, M. Moran, and R. E. Hodson. 1992. Adaptation of model genetically engineered microorganisms to lake water: Growth rate enhancements and plasmid loss. Applied and Environmental Microbiology 58:3630-3637.

Soylu, A. 1992. Heredity of male sterility in some chestnut cultivars. Acta Horticulturae 317: 181-185.

Spangenberg, G., Z. Wang, J. Nagel, and I. Potrykus. 1994. Protoplast culture and regeneration of transgenic plants in red fescue (*Festuca rubra* L.). Plant Science 97:83-94.

Spangenberg, G., Z. Y. Wang, X. L. Wu, J. Nagel, and I. Potrykus. 1995. Transgenic perennial ryegrass (*Lolium perenne*) plants from microprojectile bombardment of embryogenic suspension cells. Plant Science 108:209-217.

Spassieva, S., and J. Hille. 2002. A lesion mimic phenotype in tomato obtained by isolating and silencing an Lls1 homologue. Plant Science 162(4):543-549.

Spencer, D. 2003. Speed, containment and cost advantages of therapeutic protein production in the Lemna System™. Conference on Plant-Made Pharmaceuticals, Quebec City, Canada, March 16-19. Available online at www.cpmp2003.org/pages/en/program/program.php. Accessed April 28, 2003.

Spielman, A., J. C. Beier, and A. E. Kiszewski. 2002. Ecological and community consider-
ations in engineering arthropods to suppress vector-borne disease. Pp. 315-329 in
Genetically Engineered Organisms: Assessing Environmental and Human Health Effects.
D. K. Letourneau and B. E. Burrows, eds. Boca Raton, FL: CRC Press.

Sprang, M L., and S. E Lindow. 1981. Subcellular-localization and partial characterization of
ice nucleation activity of pseudomonas-syringae and Erwina-Herbicola. Phytopathology
7(2):256.

Srivastava, V., V. Vasil, and I. K.Vasil. 1996. Molecular characterization of the fate of
transgenes in transformed wheat (Triticum aestivum L.). Theoretical and Applied
Genetics 92:1031-1037.

St. Leger, R. J., J. Lokesh, M. J. Bidocha, and D. W. Roberts. 1996. Construction of an
improved mycoinsecticide overexpressing a toxic protease. Proceedings of the National
Academy of Sciences of the USA 93:6349-6354.

Stace, C. A. 1975. Hybridization and the Flora of the British Isles. London: Academic Press.

Staten, R., T. A. Miller, J. J. Peloquin, and E. Miller. 2001. Field release of a transgenic pink
bollworm, Pectinophora gossypiella (Lepidoptera: Gelechiidae), Dr. Robert T. Staten.
Phoenix, AZ: USDA Phoenix Plant Protection Center.

Staub J., B. Garcia, J. Graves, P. Hajdukiewicz, P. Hunter, N. Nehra, V. Paradkar, M.
Schlittler, J. Carroll, L. Spatola, D. Ward, G. Ye, and D. Russell. 2000. High yield
production of human therapeutic protein in tobacco chloroplasts. Nature Biotechnology
18(3):333-338.

Stevens, D. R., and S. Purton. 1997. Genetic engineering of eukaryotic algae: Progress and
prospects. Journal of Phycology 33:713.

Stiles, J. K., D. H. Molyneux, K. R. Wallbanks, and A. M. V. Vloedt. 1989. Effects of gamma
irradiation on the midgut ultrastructure of Glossina palpalis subspecies. Radiation
Research 118(2):353-63.

Stocker, B. A. D. 1990 Aromatic-dependent Salmonella as live vaccine presenters of foreign
epitopes as inserts in flagellin. Research in Microbiology 141:787-796.

Stokesbury, M. J., and G. L. LaCroix. 1997. High incidence of hatchery origin Atlantic
salmon in the smolt output of a Canadian river. ICES (International Council for the
Exploration of the Sea) Journal of Marine Science 54(6):1074-1081.

Stougaard J., E. Petersen, and K. Marcker. 1987. Expression of a complete soybean
leghemogloin gene in root nodules of transgenic Lotus corniculatus. Proceedings of the
National Academy of Sciences of the USA 84(16):5754-5757.

Strauss, S. H. 2003. Regulating biotechnology as though gene function mattered. BioScience
53:453-454.

Strauss, S. H., and R. Meilan. 1997. Tree genetic engineering research cooperative, report to
the U.S. Department of Energy, Office of Fuels Development. Washington, DC: Government Printing Office.

Strauss, S. H., W. H. Rottmann, A. M. Brunner, and L. A. Sheppard. 1995. Genetic engineering of reproductive sterility in forest trees. Molecular Breeding 1:5-26.

Sugiura, K., C. Inokuma, and N. Imaizumi. 1997. Transgenic creeping bentgrass (Agrostis
palustris Huds.) plants regenerated from protoplasts. Journal of Turfgrass Management
2:43-53.

Sugiura, K., C. Inokuma, N. Imaizumi, and C. Cho. 1998. Generation of herbicide resistant
creeping bentgrass (Agrostis palustris Huds.) by electroporation-mediated direct gene
transfer into protoplasts. Journal of Turfgrass Management 2:35-41.

Svab, Z., P. Hajdukiewicz, and P. Maliga. 1990. Stable transformation of plastids in higher
plants. Proceedings of the National Academy of Sciences of the USA 87:8526-8530.

Syvanen, M. 2002. Search for horizontal gene transfer from transgenic plants to microbes. Pp. 237-239 in Horizontal Gene Transfer, 2nd ed. M. Syvanen and C. I. Kado, eds. San Diego: Academic Press.

Tacon, A. G., and D. J. Brister. 2002. Organic aquaculture: Current status and future prospects. Pp. 163-176 in Organic agriculture, environment and food security, N. El-Hage Scialabba and C. Hattam, eds. Rome, Italy: Food and Agriculture Organization of the United Nations.

Talon, M., L. Zacarias, and E. Primo-Millo. 1992. Gibberellins and parthenocarpic ability in developing ovaries of seedless mandarins. Plant Physiology 99:1575-1581.

Tamura, T., C. Thibert, C. Royer, T. Kanda, E. Abraham, M. Kamba, N. Komoto, J.L. Thomas, B. Mauchamp, G. Chavancy, P. Shirk, M. Fraser, J.C. Prudhomme, P. Couble, T. Toshiki, T. Chantal, R. Corinne, K. Toshio, A. Eappen, K. Mari, K. Natuo, T. Jean-Luc, M. Bernard, C. Gerard, S. Paul, F. Malcolm, P. Jean-Claude and C. Pierre. 2000. Germline transformation of the silkworm Bombyx mori L-using a piggyBac transposon-derived vector. Nature Biotechnology 18(1):81-84.

Tang, G., B. J. Reinhart, D. P. Bartel, and P. D. Zamore. 2003. A biochemical framework for RNA silencing in plants. Genes and Development 17(1):49-63.

Tang, W., and Yingchuan Tian. 2003. Transgenic loblolly pine (*Pinus taeda* L.) plants expressing a modified -endotoxin gene of *Bacillus thuringiensis* with enhanced resistance to *Dendrolimus punctatus* Walker and *Crypyothelea formosicola* Staud. Journal of Experimental Botany 54(383):835-844.

Tave, D. 1993. Genetics for Fish Hatchery Managers. 2nd ed. New York: Van Nostrand Reinhold.

Taylor, M. R., and J. S. Tick. 2003. Post Market oversight of biotech foods. Pew Initiative on Food and Biotechnology. Washington, DC. Available online at *http://pewagbiotech.org/*

Taylor, L. H., S. M. Latham, and M. E. J. Woolhouse. 2001. Risk factors for human disease emergence. Philosophical Transactions of the Royal Societies of London. B 356 (1411):983-989.

Tenllado, F., D. Barajas, M. Vargas, F.A. Atencio, P. Gonzalez-Jara, and J.R. Diaz-Ruiz. 2003. Transient expression of homologous hairpin RNA causes interference with plant virus infection and is overcome by a virus encoded suppressor of gene silencing. Molecular Plant Microbe Interactions 16:149-158.

Thomas, D. D., C. A. Donnelly, R. J. Wood, and L. S. Alphey. 2000. Insect population control using a dominant, repressible, lethal genetic system. Science 287(5462):2474-2476.

Thomas, P. E., S. Hassan, W. K. Kaniewski, E. C. Lawson, and J. C. Zalewski. 1998. A search for evidence of virus/transgene interactions in potatoes transformed with the potato leafroll virus replicase and coat protein genes. Molecular Breeding 4:407-417.

Thompson, P. B. 1997. Food Biotechnology in Ethical Perspective. New York: Aspen Publishers.

Thompson, P. 2000. Food and agricultural biotechnology: Incorporating ethical considerations. The Canadian Biotechnology Advisory Committee Project Steering Committee on the Regulation of Genetically Modified Foods. Ottawa, ON. Available online at http://cbac-cccb.ca/epic/internet/incbac-cccb.nsf/vwapj/FoodAgric_Thompson.pdf/$FILE/FoodAgric_Thompson.pdf. Accessed December 18, 2003.

Thomson, A. J. 1999. The abundance, distribution and biology of Atlantic salmon in the North Pacific, 1991-1998. 1999 Annual Meeting of the American Fisheries Society, North Pacific International Chapter, Richmond, British Columbia, March 15-17.

Thorgaard, G. H. 1995. Biotechnology approaches to broodstock management, Pp. 76-93 in Broodstock Management and Egg and Larval Quality. N.R. Bromage and R.J. Roberts, eds. Oxford, UK: Blackwell Science.

Thorgaard, G. H., and S. K. Allen. 1992. Environmental impacts of inbred, hybrid and polyploidy aquatic species. Pp. 281-287 in Dispersal of Living organisms into Aquatic Ecosystems. A. Rosenfeld and R. Mann, eds. College Park: Maryland Sea Grant College.

Thresher, R. E., L. Hinds, P. Grewe, C. Hardy, S. Whyard, J. Patil, D. McGoldrick, and S. Vignarajan. 1999. Reversible Sterility in Animals. Australian Patent PG4884. [Note: Patent also filed in U.S.A., Canada and possibly some Asian countries]

Tiedje, J. M., R. K. Colwell, Y. L. Grossman, R. E. Hodson, R. E. Lenski, R. N. Mack, and P. J. Regal. 1989. The planned introduction of genetically engineered organisms: Ecological considerations and recommendations. Ecology 70:298-315.

Tilman, D. 1982. Resource Competition and Community Structure. Princeton, N.J: Princeton University Press.

Tolbert, K. A., M. Gopalraj, S. L. Medberry, N. E. Olszewski, and D. A. Somers. 1998. Expression of the Commelina yellow mottle virus promoter in transgenic oat. Plant Cell Reports 17(4):284-287.

Tomita, M., H. Munetsuna, T. Sato, T. Adachi, A. Hino, M. Hayashi, K. Shimizu, N. Nakamura, T. Tamura, and K. Yoshizato. 2003. Transgenic silkworms produce recombinant human type III procollagen in cocoons. Nature Biotechnology 21(1):52-56.

Tomiuk, J., T. P. Hauser, and R. Jørgensen. 2000. A- or C-chromosomes, does it matter for the transfer of transgenes from Brassica napus? Theoretical and Applied Genetics 100:750-754.

Traavik, T. 2002. Environmental risks of genetically engineered vaccines. Pp. 332-353 in Genetically Engineered Organisms: Assessing Environmental and Human Health Effects, D. K. Letourneau and B. E. Burrows, eds. Boca Raton, FL: CRC Press.

Traynor, P. L., and J. H. Westwood. 1999. Proceedings of a workshop on: Ecological effects of pest resistance genes in managed ecosystems. Blacksburg: Information Systems for Biotechnology.

Traynor, P., D. Adair, and R. Irwin. 2001. Greenhouse Research with Transgenic Plants and Microbes: A Practical Guide to Containment. Blacksburg, Va.: Information Systems for Biotechnology.

Triantaphyllidis, G. V., T. J. Abatzopoulos, R. M. Sadaltzopoulos, G. Stamou, and G. C. Kastritsis. 1993. Characterization of two new Artemia populations from two solar saltworks of Lesbos Island (Greece): Biometry, hatching characteristics and fatty acid profile. International Journal of Salt Lake Research 2(1):59-68.

Trifonova, A., and A. Atanassov. 1996. Genetic transformation of fruit and nut species. Biotechnology and Biotechnological Equipment 10:3-10.

Turner, B. J. ed. 1984. Evolutionary Genetics of Fishes. New York: Plenum Press.

Turner, P. E. 2003. Searching for the advantages of virus sex. Origins of Life and Evolution of the Biosphere 33:95-108.

Turner, P. E., and L. Chao. 1998. Sex and the evolution of intrahost competition in RNA virus. Genetics 150:523-532.

Tustolini R., and G. Cipriani. 1997. Paternal inheritance of chloroplast DNA and maternal inheritance of mitochondrial DNA in the genus Actinidia. Theoretical and Applied Genetics 94:897-903.

Tzfira, T., A. Vainstein, and A. Altmanm. 1999. rol-Gene expression in transgenic aspen (Populus tremula) plants results in accelerated growth and improved stem production index. Trees 12:49-54.

Uchtmann, D. L. 2002. Starlink—A Case Study of Agricultural Biotechnology Regulation. Drake Law Review 7:160-211.

Uchtmann, D. L., and G. C. Nelson. 2000. U.S. Regulatory Oversight of Agricultural and Food-Related Biotechnology. American Behavioural Scientist 44(1):350.

USDA (U.S. Department of Agriculture). 1997. User's Guide for Introducing Genetically Engineered Plants and Microorganisms. Technical Bulletin 1783.

USDA (U.S. Department of Agriculture). 2002. Information of field testing of pharmaceutical plants in 2002. Available online at www.aphis.usda.gov/brs/pdf/pharma_2000.pdf. Accessed November 4, 2003.

USDA (U.S. Department of Agriculture). 2003. Field testing of plants engineered to produce pharmaceutical and industrial compounds. Federal Register 68(46):11337-11340.

US EPA (U.S. Environmental Protection Agency). 1998. *Bacillus thuringiensis* subspecies *tolworthi* Cry9C protein and the genetic material necessary for its production in corn: Exemption of a requirement of a tolerance. Federal Register 63:28252.

US EPA (U.S. Environmental Protection Agency). 1999. EPA and USDA Position paper on insect resistance management on *Bt* crops. Available online at www.pestlaw.com/x/guide/1999/EPA-19990527A.html. Accessed November 4, 2003.

US EPA (U.S. Environmental Protection Agency). 2002. EPA's Regulation of *Bacillus thuringiensis* (*Bt*) Crops. Available online at www.epa.gov/oppbppd1/biopesticides/pips/regofbtcrops.htm. Accessed December 18, 2003.

US EPA (U.S. Environmental Protection Agency). 2003a. Framework for cumulative risk assessment at region 6, US EPA. Washington, DC: EPA/600/P-02/001F. Jan 1.

US EPA (U.S. Environmental Protection Agency). 2003b. Region 9 Press Release, April 23. Available online at www.pestlaw.com/x/press/2003/E09-20030423A.html. Accessed November 4, 2003.

USFWS (U.S. Fish and Wildlife Service). 2002. Snakeheads: The Newest Aquatic Invader U.S. Fish and Wildlife Service, found in Snakehead Factsheet. Available online at www.fws.gov/snakeheadfstotal.pdf. Accessed November 4, 2003.

USGS (U.S. Geological Service). 2002. Nonindigenous Aquatic Species Fact Sheet: *Trachemys scripta elegans*. Florida Integrated Science Center. Available online at *http://nas.er.usgs.gov/queries/SpFactSheet.asp?speciesID=1261*. Accessed November 4, 2003.

Utter, F. 2003. Genetic impacts of fish introductions. Pp. 357-378 in Population Genetics: Principles and Applications for Fisheries Scientists, E. M. Hallerman, ed. Bethesda, Md.: American Fisheries Society.

Vailleau, F., X. Daniel, M. Tronchet, J. L. Montillet, C. Triantaphylides, and D. Roby. 2002. A R2R3-MYB gene, AtMYB30, acts as a positive regulator of the hypersensitive cell death program in plants in response to pathogen attack. Proceedings of the National Academy of Sciences of the USA 99:10179-10184.

Valent, B., L. Farrall, and F. G. Chumley. 1991. *Magnaporthe grisea* genes for pathogenicity and virulence identified through a series of backcrosses. Genetics 127:87-101.

Van der Vloedt, A. M., and W. Klassen. 1991. The development and application of the sterile insect technique (SIT) for New World screwworm eradication. Rome, Italy: Food and Agriculture Organization World Animal Review.

Van Dijk, P., and J. van Damme. 2000. Apomixis technology and the paradox of sex. Trends in Plant Science 5:81-84.

Verma, N. K., H. K. Ziegler, B. A. D. Stocker, and G. A. Schoolnik. 1995. Induction of a cellular immune response to a defined T-cell epitope as an insert in the flagellin of a live vaccine strain of *Salmonella*. Vaccine 13:235-244.

Viard, F., J. Bernard, and B. Desplanque. 2002. Crop-weed interactions in the *Beta vulgaris* complex at a local scale: Allelic diversity and gene flow within sugar beet fields. Theoretical and Applied Genetics 104:688-697.

Visser, B., I. Van der Meer, N. Louwaars, J. Beekwilder, and D. Eaton. 2001. The impact of "Terminator" technology. Biotechnology and Development Monitor 48:9-12. Available at www.biotech-monitor.nl/4804.htm. Accessed December 18, 2003.

Wade, N. 1979. The Ultimate Experiment: Man-Made Evolution. New York: Walker and Company.

Wagner, S. 2003. Greenovation®'s Moss Bioreactor: Creating a green animal for safe cost-effective production of biopharmaceuticals. Conference on Plant-Made Pharmaceuticals, Quebec City, Canada, March 16-19. Available online at www.cpmp2003.org/pages/en/program/program.php. Accessed April 28, 2003.

Wang, R., P. Zhang, Z. Gong, and C. L. Hew. 1995. Expression of the antifreeze protein gene in transgenic goldfish *(Carassius auratus)* and its implication in cold adaptation. Molecular Marine Biology and Biotechnology 4(1):20-26.

Wang, Z., R. Wang, R. Yu, and C. Tian. 1998. Biological characteristics of polyploid shellfish. Journal of Ocean University of Qingdao/Qingdao Haiyang Daxue Xuebao 28(3):399-404.

Wang, Z., T. Takamizi, V. Iglesias, M. Ososky, J. Nagel, I. Potrykus, and G. Spangenberg. 1992. Transgenic plants of tall fescue *(Festuca arundinacea* Schreb.) obtained by direct gene transfer to protoplasts. Bio-Technology 10:691-696.

Wattendorf, R. J. 1986. Rapid identification of triploid grass carp with a coulter counter and channelyzer. The Progressive Fish Culturist 48(2):125-132.

Wattendorf, R. J., and C. Phillippy. 1996. Administration of a state permitting program. Pp. 130-176 in Managing Aquatic Vegetation with Grass Carp, a Guide for Water Resource Managers, J. R. Cassani, ed. Bethesda, Md.: American Fisheries Society.

Waters, V. L. 2001. Conjugation between bacterial and mammalian cells. Nature Genetics 29(4):375–376.

Webb, J. H., D. W. Hay, P. D. Cunningham, and A. F. Youngson. 1991. The spawning behaviour of escaped farmed and wild adult Atlantic salmon in a northern Scottish river. Aquaculture 98(1/3):97-110.

Webby, R. J., and R. G. Webster. 2003. Are we ready for pandemic influenza? Science 302:1519-1522.

Wenck, A. R., M. Quinn,R. W. Whetten, G. Pullman, and R. Sederoff. 1999. High-efficiency Agrobacterium-mediated transformation of Norway spruce *(Picea abies)* and loblolly pine *(Pinus taeda)*. Plant Molecular Biology 39:407-416.

Wesley, S. V., C. A. Helliwell, N. A. Smith, M. B. Wang, D. T. Rouse, Q. Liu, P. S. Gooding, S. P. Singh, D. Abbott, P. A. Stoutjesdijk, S. P. Robinson, A. P. Gleave, A. G. Green, and P. M. Waterhouse. 2001. Construct design for efficient, effective and high-throughput gene silencing in plants. The Plant Journal for Cell and Molecular Biology 27:581-590.

Whoriskey, F. G., and J. W. Carr. 2001. Returns of transplanted adult, escaped, cultured Atlantic salmon to the Magaguadavic River, New Brunswick. ICES (International Council for the Exploration of the Sea) Journal of Marine Sciences 58(2):504-509.

Wilkinson, M. J., I. J. Davenport, Y. M Charters, A. E. Jones, J. Allainguillaume, H. T. Butler, D. C. Mason, and A. F. Raybould. 2000. A direct regional scale estimate of transgene movement from genetically modified oilseed rape to its wild progenitors. Molecular Ecology 9:983-991.

Williams, G. C. 1975. Sex and Evolution. Princeton, NJ: Princeton University Press.

Williams, J. D., and G. K. Meffe. 2000. Invasive nonindigenous species: A major threat to freshwater biodiversity. Pp. 67-71 in Freshwater Ecoregions of North America: A Conservation Assessment, R. A. Abell., D. Olsen, E. Dinerstein, and P. Hurley, eds. Covelo, CA: Island Press.

Wilson, M., and S. E. Lindow. 1993. Effect of phenotypic plasticity on epiphytic survival and colonization by *Pseudomonas syringae*. Applied and Environmental Microbiology 59:410-416.

Wimmer, E. A. 2003. Genetically engineered, embryo-specific lethality for insect pest management. ISB (Information Systems for Biotechnology) News Report February 1-3. Available online at www.isb.vt.edu/news/2003/artspdf/mar0301.pdf. Accessed November 4, 2003.

Winn, R. N. 2001a. Bacteriophage-based transgenic fish for mutation detection. US Patent #6,307,121.

Winn, R. N. 2001b. Transgenic fish as models in environmental toxicology. ILAR (Institute for Laboratory Animal Reseach) Journal 42(4):322-329.

Winn, R. N., M. B. Norris, K. J. Brayer, C. Torres, and S. L. Muller. 2000 Detection of mutations in transgenic fish carrying a bacteriophage cII transgene target. Proceedings of the National Academy of Sciences of the USA 97(23):12655-12660.

Winn, R. N., M. B. Norris, S. L. Muller, C. Torres, and K. J. Brayer. 2001. Bacteriophage lambda and plasmid pUR288 transgenic fish models for detecting in vivo mutations. Marine Biotechnology 3 (Supplement):s185-s195.

Winn, R. N., R. J. Van Beneden, and J. G. Burkhart. 1995. Transfer, methylation and spontaneous mutation frequency of X174am3cs70 sequences in medaka (Oryzias latipes) and mummichog (Fundulus heteroclitus): Implications for gene transfer and environmental mutagenesis in aquatic species. Marine Environmental Research 40(3):247-265.

Winrock International. 2000. Transgenic crops: An environmental assessment. Henry A. Wallace Center for Agricultural and Environmental Policy at Winrock International. Available online at www.winrock.org/Transgenic.pdf. Accessed November 4, 2003.

Wipff, J., and C. Fricker. 2001. Gene flow from transgenic creeping bentgrass (Agrostis stoloifera L.) in the Willamette Valley, Oregon. International Turfgrass Society Research Journal 9:224-246.

Wolf, D. E., N. Takebayashi, and L. H. Rieseberg. 2001. Predicting the risk of extinction through hybridization. Conservation Biology 15:1039-1053.

Wolfenbarger, L. L., and P. R. Phifer. 2000. The ecological risks and benefits of genetically engineered plants. Science 290:2088-2093.

Wolters, W. R., G. S. Libey, and C. L. Chrisman. 1982. Effect of triploidy on growth and gonadal development of channel catfish. Transactions of the American Fisheries Society 111:102-105.

Woody, T. 2002. The plot to kill the carp. Wired 10(10):104-106.

Wraight, C. L., A. R. Zangerl, M. J. Carroll, and M. R. Berenbaum. 2000. Absence of toxicity of Bacillus thuringiensis pollen to black swallowtails under field conditions. Proceedings of the National Academy of Sciences of the USA 97:7700-7703.

Wright, A. D., M. B. Sampson, M. G. Neuffer, L. Michalczuk, J. P. Slovin, and J. D. Cohen. 1991. Indole-3-acetic acid biosynthesis in the mutant maize orange pericarp, a tryptophan auxotroph. Science 254:998-1000.

Wright, L. 1973. Functions. Philosophical Review 82:139-168.

Wright, S. 1969. Evolution and the genetics of populations. Volume 2. The Theory of Gene Frequencies. Chicago, IL: University of Chicago Press.

Wright, S. 1994. Molecular Politics: Developing American and British Regulatory Policy for Genetic Engineering, 1972-1982. Chicago, IL: University of Chicago Press.

Xiang, J. H., L. H. Zhou, and F. H. Li. 1999. Reproductive and genetic manipulation in Chinese shrimp, Penaeus chinensis (Osbeck 1785). Pp. 987-996 in Crustaceans and the Biodiversity Crisis. F. R. Schram and J. C. von Vaupel Klein, eds. Vol. I, Proceedings of the Fourth International Crustacean Congress, Amsterdam, The Netherlands, July 20-24, 1998. Koninklijke Brill NV, Leiden.

Xiang, J. H., L. H. Zhou, R. Liu, J. Zhu, F. Li, and X. Liu. 1993. Induction of the tetraploids of the Chinese shrimp Penaeus chinensis. Biotechnology in Agriculture 496-501.

Yamagishi, H. 1998. Distribution and allelism of restorer genes for Ogura cytoplasmic male sterility in wild and cultivated radishes. Genes and Genetic Systems 73:79-83.

Yamamoto, Y., C. Taylor, G. Acedo, and C. Cheng. 1991. Characterization of CIS acting sequences regulating root-specific gene expression in tobacco. Plant Cell 3(4):371-382.

Yang, T. W., Y. Yang, and Z. Xiong. 2000. Paternal inheritance of chloroplast DNA in interspecific hybrids in the Genus Larrea. American Journal of Botany 87:1458-2000.

Yao, J. L., and D. Cohen. 2000. Multiple gene control of plastome-genome incompatibility and plastid DNA inheritance in interspecific hybrids of Zantedeschia. Theoretical and Applied Genetics 101:400-406.

Ye, X., S. Al-Babili, A. Klöti, J. Zhang, P. Lucca, P. Beyer, and I. Potrykus. 2000. Engineering provitamin A (β-carotene) biosynthetic pathway into (carotenoid-free) rice endosperm. Science 287:303-305.

Yin, Y., L. Chen, and R. Beachy. 1997. Promoter elements required for phloem-specific gene expression from the RTBV promoter in rice. The Plant Journal for Cell and Molecular Biology 12:1179-1188.

Youngson, A. F., J. H. Webb, J. C. MacLean, and B. M. Whyte. 1997. Short communication: Frequency of occurrence of reared Atlantic salmon in Scottish salmon fisheries. ICES (International Council for the Exploration of the Sea) Journal of Marine Sciences 54(6):1216-1220.

Zaslavskaia, L. A., J. C. Lippmeier, C. Shih, D. Ehrhardt, A. R. Grossman, and K. E. Apt. 2001. Trophic conversion of an obligate photoautotrophic organism through metabolic engineering. Science 292:2073-2075.

Zhan X., H. Wu, and A. Cheung. 1996. Nuclear male sterility induced by pollen-specific expression of a ribonuclease. Sexual Plant Reproduction 9(1):35-43.

Zhang, P., Y. Xu, Z. Liu, Y. Xiang, S. Du, and C. L. Hew. 1998. Gene transfer in red sea bream (Pagrosomus Major). Pp. 15-18 in New Developments in Marine Biotechnology, Y. LeGal and H. O. Halvorson, eds. New York: Plenum Press.

Zhang, X. C., Y. X. Mao, G. G. Wang, B. H. Zhang, G. P. Yang, and Z. H. Sui. 2001. Preliminary studies on the genetic transformation of Spirulina platensis. Pp. 263-269 in Algae and their Biotechnological Potential: Proceedings of the 4th Asia-Pacific Conference on Algal Biotechnology, 3-6 July 2000 in Hong Kong. F. Chen, and Y. Jiang, eds. Boston, MA: Kluwer Academic Publishers.

Zhao, Z., Y. Cao, M. Li, and A. Meng. 2001. Double-stranded RNA injection produces nonspecific defects in zebrafish. Developmental Biology 229(1):215-223.

Zhivotovsky, B. 2002. From the nematode and mammals back to the pine tree: On the diversity and evolution of programmed cell death. Cell Death and Differentiation 9:867-869.

Zhong, H., C. A. Liu, J. Vargas, D. Penner, and M. Sticklen. 1998. Simultaneous control of weeds, dollar spot and brownpatch diseases in transgenic creeping bentgrass. Pp. 203-210 in Turfgrass Biotechnology: Cell and Molecular Genetic Approaches to Turfgrass Improvement, M. Sticklen, and M. Kenna, eds. Chelsea, Michigan: Ann Arbor Press, Inc.

Zhong, H., F. Teymouri, B. Chapman, S. Maqbool, R. Sabzikar, Y. El-Maghraby, B. Dale, and M. B. Sticklen. 2003. The pea (Pisum sativum L.) rbcS transit peptide directs the Alcaligenes eutrophus polyhydroxybutyrate enzymes into the maize (Zea mays L.) chloroplasts. Plant Science 165(3):455-462.

Zhong, H., M. Bolyard, C. Srinivasan, and M. Sticklen. 1993. Transgenic plants of turfgrass (Agrostis palustris huds.) from microprojectile bombardment of embryogenic callus. Plant Cell Reproduction 13:1-6.

Zhou, M. F. 2002. Chromosome set instability in 1-2 year old triploid Crassostrea ariakensis in multiple environments. M. A. Thesis. Gloucester Point, VA: Virginia Institute of Marine Science.

Zhu, T., H. L. Mogensen, and S. E. Smith. 1993. Quantitative, three-dimensional analysis of alfalfa egg cells in two genotypes: Implications for biparental plastid inheritance. Planta 190:143-150.

Zhu, Z. 2001. Institute of Hydrobiology, Chinese Academy of Sciences, People's Republic of China; personal communication, November 20.

Zund, P., and G. Lebek. 1980. Generation time-prolonging R plasmids: Correlation between increases in the generation time of *Escherichia coli* caused by R plasmids and their molecular size. Plasmid 3:65-69.

Zuo, J., Q. W. Niu, S. G. Moller, and N. H. Chua. 2001. Chemical-Regulated, Site-Specific DNA Excision in Transgenic Plants. Nature Biotechnology 19:157-161.

About the Authors

T. Kent Kirk, *Chair,* is Professor Emeritus in the Department of Bacteriology, University of Wisconsin, and is retired from the U.S. Department of Agriculture Forest Products Laboratory in Madison. He is best known for his research on the microbiological degradation of lignin. He and coworkers discovered the pathways and isolated the oxidative enzyme system by which this degradation is achieved in nature. Dr. Kirk also has made contributions to the industrial application of fungi and their enzymes, primarily in the pulp and paper industry and in bioremediation of soil and water. He has served as a member and as chair of the National Research Council's Board on Agriculture and Natural Resources, and has served on other NRC committees. He was the 1985 recipient of the Marcus Wallenberg Prize for forest-related research, and was elected to the National Academy of Sciences in 1988. Dr. Kirk received his Ph.D. in Biochemistry and Plant Pathology from North Carolina State University in 1968, and completed postdoctoral training in organic and polymer chemistry in Sweden.

John E. Carlson is Associate Professor of Molecular Genetics and Director of the Schatz Center for Tree Molecular Genetics in the School of Forest Resources at The Pennsylvania State University. His expertise is in molecular genetics and biotechnology, and he works primarily with forest species. Dr. Carlson's research includes studies on genetic diversity in natural populations, the structure of plant genomes, and the micropropagation of Christmas trees. Dr. Carlson's group also conducts research on the molecular basis and modification of lignin synthesis in trees and the response of trees to environmental stress. Dr. Carlson currently is Chair of the International Union of Forest Research Organizations study group on Genetics of *Quercus* (oak). He also has served as a scientific advisor for the Association of Southeast Asian Nations Forest Tree Center in Bangkok, Thailand, as well as with the National Laboratory of Forest Ecology at Harbin University, China. He earned his Ph.D. in genetics from the University of Illinois, Urbana Champaign in 1983.

Norman Ellstrand is Professor of Genetics in the Department of Botany and Plant Sciences, University of California, Riverside. He also is Director of its Biotechnology Impacts Center. Dr. Ellstrand's research focuses on applied plant population genetics with current emphasis on the consequences of gene flow from domesticated plants to their wild relatives, including the hybridization of transgenic crops with wild relatives. He has participated in several government and National Research Council meetings concerning genetically engineered organisms, including the 2002 NRC Committee on Environmental Effects Associated with Commercialization of Transgenic Plants. His honors range from a Fulbright Fellowship to Sweden to being chosen the first Distinguished Speaker at the United Kingdom's Ecological Genetics Group meeting. He received his Ph.D. in biology from the University of Texas, Austin in 1978.

Anne R. Kapuscinski is Professor in the Department of Fisheries, Wildlife, and Conservation Biology, University of Minnesota. She is recognized for her expertise in fish conservation genetics, biosafety assessment of genetically engineered organisms, and impacts of aquaculture. Dr. Kapuscinski's research addresses effects of artificially propagated fish on the fitness and genetic diversity of wild relatives, measuring these effects for hatchery-released, farm-escaped and transgenic fish. Her laboratory also is researching temporal genetic trends in wild fish populations and organic aquaculture. She received a U.S. Department of Agriculture Secretary Honor Award (1997) and a Pew Fellowship in Marine Conservation (2001) for her linkage of science to public policy regarding aquatic biotechnology and fish genetic conservation. Dr. Kapuscinski is the founding Director of the Institute for Social, Economic, and Ecological Sustainability at the University of Minnesota, where she leads programs that engage government, academia, and the nongovernmental sector in sustainability and biotechnology issues. She served on the NRC Committee on Atlantic Salmon in Maine, the NRC Committee on Protection and Management of Pacific Northwest Anadromous Salmonids, and the independent Scientists' Working Group on Biosafety. Most recently, Dr. Kapuscinski was appointed to the Global Environmental Facility's Scientific and Technical Advisory Panel as a biosafety advisor, the Food and Drug Administration's Food Safety Committee, Subcommittee on

Biotechnology, a Consultative Group on International Agricultural Research Study Panel on safe uses of gene technology, and a Food and Agriculture Organization/World Health Organization Expert Consultation on food safety of genetically engineered fish. Kapuscinski earned her Ph.D. in fisheries sciences in 1984 from Oregon State University.

Thomas A. Lumpkin is Director General of the Asian Vegetable Research and Development Center in Taiwan. He previously served as Chair of the Department of Crop and Soil Sciences at Washington State University, and Professor of Agronomy and of Asian Studies. He was Co-Chair of the Washington State University Advisory Council for International Affairs, and his research has concentrated on East Asian agriculture with a focus on China and Japan. Dr. Lumpkin has held a number of international consultancies concerning the agricultural sciences, including the United Nations Food and Agriculture Organization, and he is a former National Academy of Sciences Research Scholar in China. He has published numerous peer-reviewed articles, book chapters, and several books ranging from topics such as soybean genetics to management practices on the yield of paddy rice. Dr. Lumpkin holds a Ph.D. in agronomy from the University of Hawaii in 1983.

David C. Magnus is Associate Professor of Pediatrics (Medical Genetics) and Co-Director of the Stanford Center for Biomedical Ethics. He is recognized for his expertise in biology and bioethics, especially as applied to genetic technology and agricultural biotechnology. Dr. Magnus has published widely on the history and philosophy of biology and bioethics. He serves as Associate Editor of the American Journal of Bioethics and as an editorial advisor for the *Encyclopedia of Life Sciences*. Dr. Magnus is a member of the U.S. Secretary of Agriculture's Advisory Committee on Biotechnology and Agriculture in the 21st Century. He has acted as a consultant for the National Conference of State Legislators on cloning, and as an expert consultant for the World Bank on food security and biotechnology. Dr. Magnus earned a Ph.D. from Stanford University in 1989.

Daniel B. Magraw, Jr. is President of the Center for International Environmental Law (CIEL) in Washington, D.C. His expertise is in international law and public policy, including the regulation of genetically engineered organisms that are released into the environment. From 1992 to 2001, he served the U.S. Environmental Protection Agency as Associate General Counsel and Director of the International Environmental Law Office, before joining CIEL, a public interest law organization that uses principles of ecology and justice to strengthen international environmental law. Mr. Magraw has published numerous articles and books concerning public and private international law, and he has served as Chair of the Section of International Law and Practice of the American Bar Association. He holds a J.D. from the University of California, Berkeley, School of Law (1976), a BA (high honors in economics, 1968) from Harvard University, and served as a Professor of Law at the University of Colorado from 1983 to 1992.

Eugene W. Nester, Professor of Microbiology at the University of Washington, is recognized for his knowledge of molecular genetics. Nester is known for his creative and interdisciplinary approach to solving problems in biochemistry and molecular genetics. He was the first to demonstrate *Agrobacterium* Ti plasmid gene transfer into plants, and has contributed much to our current understanding of the mechanisms for incorporating foreign genes into plants. Given that *Agrobacterium* is the main mechanism used in biotechnology for introducing recombinant DNA into target organisms, Nester's research has been essential for optimizing this system of gene transfer for research and commercial purposes. In addition to earning several awards, such as the Australia Prize and the Cetus Award in Biotechnology, Nester was elected to the National Academy of Sciences in 1994. He has served on several National Research Council committees, most notably on the 1989 Committee on Scientific Introduction of Genetically Modified Microorganisms and Plants into the Environment.

John J. Peloquin is Group Leader for Protein Chemistry for the American Protein Corporation in Ames, Iowa. Previously, Dr. Peloquin was an Assistant Researcher V in the Departments of Entomology and Biochemistry at University of California, Riverside. Currently, his research interests include the identification, purification and industrial production of functional proteins from complex sources in addition to his previous study of insect symbiotic bacteria, and the genetic transformation of insects for eventual field release. He has authored or co-authored numerous peer-reviewed publications and book chapters concerning symbiotic extracellular bacteria, pink bollworm, transgenic insects, and other topics related to parasitology and heterologous gene expression. Dr. Peloquin has served as a consultant to the United Nations Food and Agriculture Organization, as well as a study group member of the First International Workshop on Comparative Insect Genomics. He has worked to develop patents for inventions such as a pink bollworm expression system, as well as an electroelution device for nucleic acids and proteins. Dr. Peloquin earned his Ph.D. in vector biology from Texas A&M University in 1985.

Allison A. Snow is Professor of Evolution, Ecology, and Organismal Biology at The Ohio State University. She is noted for her expertise in the microevolutionary processes of plant populations, including breeding systems, pollination ecology, and conservation biology. Currently, Dr. Snow's research focuses on whether crop-to-wild gene flow can lead to rapid evolution in weeds. She has published widely in numerous peer-reviewed journals, in addition to having published several technical reports and book chapters on topics such as transgenic plants, pollination ecology, and gene flow. Dr. Snow is a member of the National Research Council Committee on Agricultural Biotechnology, Health, and the Environment, and she served on the 1999 NRC Committee on Genetically Modified Pest-Protected Plants. She is also President-Elect of the Botanical Society of America, and is an Associate Editor for Environmental Biosafety Research. Dr. Snow received her Ph.D. in botany from the University of Massachusetts in 1982.

Mariam B. Sticklen is a Professor in the Department of Crop and Soil Sciences at Michigan State University. She is known for her expertise in genetic engineering of cereals and turfgrasses. Dr. Sticklen has been extensively engaged in the genetic improvement of agricultural crops of developing countries, reducing applications of hazardous chemicals by gene discovery and cloning, and genetic engineering of turfgrasses and other gramineous plants, and production of ethanol and other biobased industrial products (including biodegradable plastic) in corn through genetic engineering. She has been a member of the Board of Trustees for the International Crops Research Institute for the Semi-Arid Tropics in India and Africa for six years. In 1991, she established a project titled "Agricultural Biotechnology for Sustainable Productivity," through the United States Agency for International Development. Dr. Sticklen has received many awards for her contributions, including the Michigan State University's Ralph Smuckler International Award. She currently owns six patents and has published two books and many peer-reviewed articles. Dr. Sticklen received a Ph.D. in horticulture from The Ohio State University in 1981.

Paul E. Turner is Assistant Professor in the Department of Ecology and Evolutionary Biology at Yale University. Dr. Turner's research interests include virus ecology, host–parasite interactions, the evolution of infectious disease, and the evolution of sex. His laboratory is using microorganisms as model systems to address hypotheses in ecological and evolutionary theory. His other research interests include the role of sex and its consequences in virus evolution, game theory and virus interactions, evolution of plasmid transmission, and virus mediation of host apoptosis. Dr. Turner has published numerous peer-reviewed articles on topics in these areas, and he has spoken at several professional workshops and symposia in the U.S. and abroad. He earned his Ph.D. in zoology from Michigan State University, with a certificate in ecology and evolution, in 1995.

Board on Agriculture and Natural Resources

PUBLICATIONS

Policy and Resources

Agricultural Biotechnology and the Poor: Proceedings of an International Conference (2000)

Agricultural Biotechnology: Strategies for National Competitiveness (1987)

Agriculture and the Undergraduate: Proceedings (1992)

Agriculture's Role in K-12 Education: A Forum on the National Science Education Standards (1998)

Air Emissions from Animal Feeding Operations: Current Knowledge, Future Needs (2003)

Alternative Agriculture (1989)

Animal Biotechnology: Science-Based Concerns (2002)

Brucellosis in the Greater Yellowstone Area (1998)

Colleges of Agriculture at the Land Grant Universities: Public Service and Public Policy (1996)

Colleges of Agriculture at the Land Grant Universities: A Profile (1995)

Countering Agricultural Bioterrorism (2003)

Designing an Agricultural Genome Program (1998)

Designing Foods: Animal Product Options in the Marketplace (1988)

Diagnosis and Control of Johne's Disease (2003)

Ecological Monitoring of Genetically Modified Crops (2001)

241

Ecologically Based Pest Management: New Solutions for a New Century (1996)

Emerging Animal Diseases—Global Markets, Global Safety: A Workshop Summary (2002)

Ensuring Safe Food: From Production to Consumption (1998)

Environmental Effects of Transgenic Plants: The Scope and Adequacy of Regulation (2002)

Exploring Horizons for Domestic Animal Genomics: Workshop Summary (2002)

Forested Landscapes in Perspective: Prospects and Opportunities for Sustainable Management of America's Nonfederal Forests (1997)

Frontiers in Agricultural Research: Food, Health, Environment, and Communities (2003)

Future Role of Pesticides for U.S. Agriculture (2000)

Genetic Engineering of Plants: Agricultural Research Opportunities and Policy Concerns (1984)

Genetically Modified Pest-Protected Plants: Science and Regulation (2000)

Incorporating Science, Economics, and Sociology in Developing Sanitary and Phytosanitary Standards in International Trade: Proceedings of a Conference (2000)

Investing in Research: A Proposal to Strengthen the Agricultural, Food, and Environmental System (1989)

Investing in the National Research Initiative: An Update of the Competitive Grants Program in the U.S. Department of Agriculture (1994)

Managing Global Genetic Resources: Agricultural Crop Issues and Policies (1993)

Managing Global Genetic Resources: Forest Trees (1991)

Managing Global Genetic Resources: Livestock (1993)

Managing Global Genetic Resources: The U.S. National Plant Germplasm System (1991)

National Capacity in Forestry Research (2002)

National Research Initiative: A Vital Competitive Grants Program in Food, Fiber, and Natural Resources Research (2000)

New Directions for Biosciences Research in Agriculture: High-Reward Opportunities (1985)

Pesticide Resistance: Strategies and Tactics for Management (1986)

Pesticides and Groundwater Quality: Issues and Problems in Four States (1986)

Pesticides in the Diets of Infants and Children (1993)

Precision Agriculture in the 21st Century: Geospatial and Information Technologies in Crop Management (1997)

Predicting Invasions of Nonindigenous Plants and Plant Pests (2002)
Professional Societies and Ecologically Based Pest Management (2000)
Rangeland Health: New Methods to Classify, Inventory, and Monitor
    Rangelands (1994)
Regulating Pesticides in Food: The Delaney Paradox (1987)
Resource Management (1991)
The Scientific Basis for Estimating Air Emissions from Animal Feeding
    Operations: Interim Report (2002)
Soil and Water Quality: An Agenda for Agriculture (1993)
Soil Conservation: Assessing the National Resources Inventory, Volume
    1 (1986); Volume 2 (1986)
Sustainable Agriculture and the Environment in the Humid Tropics
    (1993)
Sustainable Agriculture Research and Education in the Field: A
    Proceedings (1991)
Toward Sustainability: A Plan for Collaborative Research on Agriculture
    and Natural Resource Management (1991)
Understanding Agriculture: New Directions for Education (1988)
The Use of Drugs in Food Animals: Benefits and Risks (1999)
Water Transfers in the West: Efficiency, Equity, and the Environment
    (1992)
Wood in Our Future: The Role of Life Cycle Analysis (1997)

Nutrient Requirements of Domestic Animals Series and Related Titles

Air Emissions from Animal Feeding Operations: Current Knowledge,
    Future Needs (2003)
Building a North American Feed Information System (1995) (available
    from the Board on Agriculture)
Metabolic Modifiers: Effects on the Nutrient Requirements of Food-
    Producing Animals (1994)
Nutrient Requirements of Dogs and Cats (2003)
Nutrient Requirements of Beef Cattle, Seventh Revised Edition, Update
    (2000)
Nutrient Requirements of Cats, Revised Edition (1986)
Nutrient Requirements of Dairy Cattle, Seventh Revised Edition (2001)
Nutrient Requirements of Dogs, Revised Edition (1985)
Nutrient Requirements of Fish (1993)
Nutrient Requirements of Horses, Fifth Revised Edition (1989)
Nutrient Requirements of Laboratory Animals, Fourth Revised Edition
    (1995)
Nutrient Requirements of Nonhuman Primates, Second Revised Edition
    (2003)

Nutrient Requirements of Poultry, Ninth Revised Edition (1994)
Nutrient Requirements of Sheep, Sixth Revised Edition (1985)
Nutrient Requirements of Swine, Tenth Revised Edition (1998)
Predicting Feed Intake of Food-Producing Animals (1986)
Role of Chromium in Animal Nutrition (1997)
Ruminant Nitrogen Uses (1985)
Scientific Advances in Animal Nutrition: Promise for the New Century
     (2001)
Vitamin Tolerance of Animals (1987)

Further information, additional titles (prior to 1984), and prices are available from the National Academies Press, 500 Fifth Street, NW, Washington, D.C. 20001, 202-334-3313 (information only). To order any of the titles you see above, visit the National Academies Press bookstore at *http:// www.nap.edu/bookstore.*

Index

A

Abandoned methods, 152–153
Action Group on Erosion, Technology, and
    Concentration, 25
Adequacy of confinement, 54–55
Advisory Board for the Research Councils,
    20
Agricultural Research Service, 25
Allopolyploids, chromosomal locations in,
    88–90
Animal and Plant Health Inspection Service
    (APHIS), 24, 36, 54, 56, 115
Animal Biotechnology: Science-based
    Concerns, 35
Animals
    bioconfinement of, 130–158
    methods of bioconfinement, 4–5
Antifertility genes, 23
APHIS. *See* Animal and Plant Health
    Inspection Service
Apomixis (asexually produced seeds), 81–83
    strengths, 82
    weaknesses, 83
Aqua Bounty Farms, 137–138
Artificially induced transgene expression,
    84–85
    strengths, 85
    weaknesses, 85

Asexually produced seeds, 81–83
Ashby Committee, 20
Auxotrophy, 111–112
Aventis CropScience, 24, 34

B

Bacteria
    bioconfinement of, 169–179
    displacement of indigenous
        populations, 164–165
    horizontal genetic transfer into local
        populations, 167–168
    invasion into indigenous populations,
        162–164
    phenotypic handicapping, 170–172
    suicide genes, 173–174
BANR. *See* Board on Agriculture and
    Natural Resources
Barriers
    physical, 16
    physiochemical, 16–17
Berg, Paul, 19
Bio-barcodes, 10
Bioconfinement, 180–185
    adequacy of, 54–55, 184
    case-by-case evaluation, 180–181
    changes of efficacy with scale, 184

compliance, 190–191
cost of compliance, 192–193
defined, 15–16
defining risk, 30–35
delaying the evolution of resistance,
    52–53
early evaluation, 181–182
effects on nontarget species, 52
ensuring efficacy of, 6–10
experimental information on efficacy,
    183–184
food safety and other issues, 53
history of, 19–25
and human error, 191–192
increases in spatial and temporal
    scale, 193
increasing the efficacy of, 189–193
natural events, 192
necessary for trees, 105–106
need for, 55–56
need for preventive actions, 53–54
options based on technology and
    gene-specific compounds, 185
predicting the consequences of failure,
    56–58
private litigation, 193
rationale for, 3–4
redundancy, 182–183
transparency and public participation,
    189–190
unacceptability of some methods
    under some circumstances, 185
when and why to consider, 29–64
who decides, 58–64
Bioconfinement concerns, 35–52
about field-released GEOs, 48–52
gene dispersal and persistence, 38–48
potential effects, 36–38
Bioconfinement failure, 194–195
detecting and mitigating, 10–11
failed or inappropriate methods of
    microbial, 176
managing, 124–129
predicting the consequences of, 56–58
StarLink corn, 34
Bioconfinement methods, 4
in animals, 4–5
in microbes, 5–6
in plants, 4, 66–68
social acceptability of, 25–28
for transgenic turfgrasses, 120–121

unacceptability of some methods
    under some circumstances, 185
verification, monitoring, and the
    efficacy of confinement, 191
Bioconfinement of animals, 130–158
bioconfinement of fish and shellfish,
    132–153
bioconfinement of insects, 153–158
Bioconfinement of bacteria, viruses, and
    fungi, 160–179
displacement of indigenous
    populations, 164–165
ecosystem and population effects,
    177–178
effectiveness of methods at different
    temporal and spatial scales, 176–
    177
failed or inappropriate methods of
    microbial bioconfinement, 176
fitness reduction, 169–176
horizontal genetic transfer into local
    populations, 165–169
invasion into indigenous populations,
    161–164
microalgae, 179
monitoring, detection, and culling—
    needs, feasibility, and realities,
    178
Bioconfinement of fish and shellfish, 132–
    153
abandoned and inappropriate
    methods, 152–153
combining triploidization with
    interspecific hybrids, 152
disruption of sexual reproduction,
    132–145
gene blocking and gene knockout,
    145–151
naturally sterile interspecific hybrids,
    151–152
Bioconfinement of insects, 153–158
ecological characteristics of
    production site, 157–158
fitness reduction and regulation of
    gene expression, 158
sterile insect technique, 153–156
transgenic sterile insects, 156–157
Bioconfinement of plants, 65–129
effectiveness at different spatial and
    temporal scales, 122–124
genetically engineered trees, 98–114

methods of bioconfinement, 65–97
monitoring and managing
    confinement failure, 124–129
Bioconfinement of pollen-mediated spread
        of transgenes, 76–83
    apomixis (asexually produced seeds),
        81–83
    cleistogamy (closed flowers), 81
    nontransgenic male sterility, 76–77
    transgenes in chloroplast DNA, 79–81
    transgenic male sterility, 77–78
Bioconfinement of transgenes in trees,
        outlook for, 113–114
Bioconfinement of trees, 98, 104–106
Bioconfinement redundancy, 17
Biological and operational considerations
        for bioconfinement, 180–198
    bioconfinement failure, 194–195
    execution of confinement, 185–193
    international aspects, 193–194
    looking to the future: strategic public
        investment in bioconfinement
        research, 195–198
"Biological containment," 20
BLS. *See* Board on Life Sciences
Board on Agriculture and Natural
        Resources (BANR), 2
Board on Life Sciences (BLS), 2
Brenner, Sydney, 20

C

Case-by-case evaluation, 180–181
    recommendation to evaluate each
        GEO separately, 181
CBI. *See* Confidential business information
Changes of efficacy with scale, 184
    recommendation to assess
        bioconfinement techniques with
        reference to temporal and spatial
        scales of field release, 184
*Channeling, Identity Preservation and the
    Value Chain: Lessons from the
    Recent Problems with StarLink
    Corn,* 35
χ1776, 172
Chloroplast-targeting gene expression, 93
Choices
    of alternative organisms or
        "abstinence," 96–97
    not to proceed, 97

Chromosomal locations in allopolyploids,
        88–90
    weaknesses, 90
Citizen suits, to enforce environmental
        laws, 63
Clean Water Act, 63
Cleistogamy (closed flowers), 81
    strengths, 81
    weaknesses, 81
Committee on Biological Confinement of
        Genetically Engineered
        Organisms, 2, 30, 130
Community effects, 126–129
Competition, heightened, 141
Compliance
    cost of, 192–193
    and the efficacy of confinement, 190–
        191
Concerns, 35–52
    about field-released GEOs, 48–52
    about gene dispersal and persistence,
        38–48
    about potential effects, 36–38
    See Bioconfinement concerns
Confidential business information (CBI), 62
Confinement. *See* Bioconfinement
Consequentialism, and public acceptance,
        27–28
Constraints, 110–113
    fitness handicaps, 112
    gene silencing, 111–112
    plastid engineering, 113
    sterility, 110–111
    tissue-specific expression, 112–113
    triploidy, 111
Consultative Group on International
        Agricultural Research, 26
Conway, Gordon, 26
Cooperative Research and Development
        Agreement (CRADA), 25
Coordinated Framework for the Regulation
        of Biotechnology, 1, 59–60
Cost of compliance, and the efficacy of
        confinement, 192–193
CRADA. *See* Cooperative Research and
        Development Agreement
Cross-incompatibility, 88
    strengths, 88
    weaknesses, 88
Cry9C gene, 24, 56
Culling, needs, feasibility, and realities, 178

D

Decision making, 185–186
Decisions about when and why to consider
        bioconfinement, 58–64
    citizen suits to enforce environmental
        laws, 63
    government, 59–63
    industry, 58–59
    insurance companies, 59
    private action for damage, 63–64
Default gene blocking by interference RNA
        and exogenous rescue, 149
    strengths, 149
    weaknesses, 149
Delta and Pine Land Company, 23, 25
Detection, needs, feasibility, and realities,
        178
Detection technology, 187–189
    recommendation to develop easily
        identifiable markers, 187–189
Dispersal biology of organisms targeted for
        genetic engineering and release,
        recommendation to support
        additional scientific research to
        develop better understanding of,
        197
Dispersal of transgenes, concerns about,
        38, 46–47
Displacement of indigenous populations,
        164–165
    bacteria, 164–165
    viruses, 164
Disruption of sexual reproduction, 132–
        145

E

Early cell division, normal steps in, 134
Early evaluation, 181–182
    recommendation to consider the need
        for bioconfinement early in the
        development of a GEO or its
        products, 182
Ecological characteristics, of production
        sites, 157–158
Ecological consequences, of large-scale use
        of bioconfinement, 11–13
Economic factors influencing application
        and regulation of particular
        techniques, 197

Ecosystem effects, 126–129, 177–178
Effectiveness of methods, at different
        temporal and spatial scales, 122–
        124, 176–177
ELISAs. See Enzyme-linked immunosorbent
        assays
Environmental effects of confinement
        failure, recommendation to
        support additional scientific
        research to assess, 196
Environmental Effects of Transgenic
        Plants, 31
Environmental laws, citizen suits to
        enforce, 63
Environmentally sound bioconfinement,
        recommendation to support
        additional scientific research to
        develop, 196
Enzyme-linked immunosorbent assays
        (ELISAs), 125
Eradication or control of escaped
        organisms, 189
Ethical factors influencing application
        and regulation of particular
        techniques, 197
Evaluation, early, 181–182
Evolutionary persistence of transgenes,
        concerns about, 47–48
Excision of transgenes before reproduction,
        84
    strengths, 84
    weaknesses, 84
Execution of confinement, 185–193
    decision making, 185–186
    eradication or control of escaped
        organisms, 189
    increasing the efficacy of confinement,
        189–193
    integrated confinement system, 186–
        187
    monitoring and detection technology,
        187–189
    research, 186
Experimental information on efficacy, 183–
        184
    recommendation to compare the novel
        genotype with its progenitor
        before field release, 183–184
    recommendation to test confinement
        techniques before putting them
        into application, 183

Externally administered gene-specific
  compounds, 149–150
  strengths and weaknesses, 150
Extinction of wild taxa, concerns about,
  50–51

F

Failure. *See* Bioconfinement failure
Farmer reluctance, 141
Federal Food, Drug, and Cosmetics Act
  (FFDCA), 61
Federal Insecticide, Fungicide, and
  Rodenticide Act (FIFRA), 62
Festuca arundinacea, 119
Field release choice, 96–97
Field-released GEOs, 48–52
  extinction of wild taxa, 50–51
  gene flow to other domesticated
    organisms, 50–51
  weediness or invasiveness, 49–50
*Field Testing Genetically Modified
  Organisms,* 160
First International Symposium on
  Sustainable Fish Farming, 23
Fish, bioconfinement of, 132–153
Fitness handicaps, 91–93, 112
  strengths, 92
  weaknesses, 92–93
Fitness reduction, 169–176
  χ1776, 172
  phenotypic handicapping, 169–173
  and regulation of gene expression,
    158
  strengths, 91
  suicide genes, 173–176
  in transgenic crop-wild progeny, 90–
    91
  weaknesses, 91
Flavr Savr™ tomato, 36
Flower- and fruit-specific gene expression,
  94–95
Food safety issues, 53
Fungi
  bioconfinement of, 169–179
  horizontal genetic transfer into local
    populations, 168–169
  invasion into indigenous populations,
    161–162
  phenotypic handicapping, 172–173
  suicide genes, 174–175

G

Gamete fertilization, normal steps in, 134
GE. *See* Genetically engineered species
Gene blocking, 145–151
  default gene blocking by interference
    RNA and exogenous rescue, 149
  externally administered gene-specific
    compounds, 149–150
  genetic bioconfinement strategies for
    fish, 147
  inducible transgenic gene blocking or
    misexpression, 146–148
Gene dispersal and persistence, 38–48
  evolutionary persistence of transgenes,
    47–48
  genetically engineered organisms, 39
  how transgenes disperse, 38, 46–47
Gene expression
  flower- and fruit-specific, 94–95
  roots and tuber-specific, 93–94
Gene flow
  from genetically engineered
    organisms, 4
  to other domesticated organisms,
    concerns about, 50–51
  potential for, 118–120
Gene knockout, 150–151
  strengths and weaknesses, 150–151
Gene silencing, 103, 111–112
Gene Tools, 150
Genetic bioconfinement, strategies for fish,
  147
"Genetic use restriction technologies"
  (GURTs), 26
Genetically Engineered Food Alert, 24
Genetically engineered species, 17
Genetically engineered organisms, 1
  concerns about, 39
  current and future, 36–38
  defined, 14–15
  finfish, 39–41
  gene flow from, 4
  insects, 45
  marine microorganisms, 42
  marine plants, 41–42
  microbes, 44–45
  mollusks, 41
  terrestrial plants, 42–44
Genetically engineered trees, 98–114
  bioconfinement of, 98, 104–106

future necessity of biological
confinement for trees, 105–106
options and constraints, 110–113
outlook for bioconfinement of
transgenes in trees, 113–114
risks of most concern with trees, 106–
110
stability of transgenic confinement,
102–103
Genetically engineered turfgrasses, 115–
120
bioconfinement methods for, 120–121
difficulty of confinement, 120
potential for gene flow, 118–120
wild hybrid, *Festuca arundinacea* and
Lolium multiflorum Lam., 119
Geographic Information System
technology, 183
GEOs. *See* Genetically engineered
organisms
GFP. *See* Green fluorescent protein (GFP)
Government, decision-making about
bioconfinement, 59–63
Green fluorescent protein (GFP), 114, 156,
162
Green-specific (chloroplast-targeting) gene
expression, 93
GURTs. *See* "Genetic use restriction
technologies"

H

Horizontal genetic transfer, 38, 165–169
bacteria, 167–168
fungi, 168–169
viruses, 166
HR. *See* Hypersensitive response
Human error, 191–192
and the efficacy of confinement, 191–
192
recommendation to take into account
when determining
bioconfinement methods and
evaluating their efficacy, 192
Hybrids, interspecific, 69
Hypersensitive response (HR), 75

I

ICS. *See* Integrated confinement system

Inducible transgenic gene blocking or
misexpression, 146–148
strengths, 147–148
weaknesses, 148
Industry, decision-making about
bioconfinement, 58–59
Influences on the application and
regulation of particular
techniques, recommendation to
support additional scientific
research to identify, 197
Information, right of the public to, 10
Insects
bioconfinement of, 153–158
subjected to the sterile insect
technique, 155
Instability of transgene expression, 110
Insurance companies, decision-making
about bioconfinement, 59
Integrated confinement system (ICS), 8, 34,
186–187
recommendation to use an ICS
approach for GEOs that warrant
confinement, 187
Intellectual property rights, 26
International aspects, 19, 193–194
recommendation to consider potential
effects of a confinement failure
on other nations, 194
recommendation to pursue
international cooperation to
adequately manage confinement
of GEOs, 194
Interspecific hybrids, 69
strengths, 69
weaknesses, 69
Invasion biology, recommendation to
support additional scientific
research to develop better
understanding of, 197–198
Invasion into indigenous populations, 161–
164
bacteria, 162–164
fungi, 161–162
viruses, 161
Invasiveness, concerns about, 49–50

L

Leakage, 80

Legal factors, influencing application and regulation of particular techniques, 197
Lolium multiflorum Lam., 119

M

Managing confinement failure, 124–129
  population, community, and ecosystem effects, 126–129
Methods of bioconfinement. *See* Bioconfinement methods
Microalgae, 179
Microbes, 5–6
Monitoring
  difficulty of, 11
  needs, feasibility, and realities, 178
Monitoring confinement failure, 124–129
Monitoring technology, 187–189
Monsanto, 23, 26, 115
Mortality of vegetative propagules, 75–76
  strengths, 76
  weaknesses, 76
Mosaic individuals, 140

N

National Academy of Sciences (NAS), 20, 160
National Endangered Species Act, 63
National Environmental Policy Act (NEPA), 22, 62–63
National Institutes of Health (NIH), 20–22
  Recombinant DNA Advisory Committee, 21–22
National Marine Fisheries Service, 62
National Research Council, 31
  Board on Agriculture and Natural Resources, 2
  Board on Life Sciences, 2
Natural events, and the efficacy of confinement, 192
Naturally sterile interspecific hybrids, 151–152
  strengths, 151
  weaknesses, 151–152
NEPA. *See* National Environmental Policy Act
NGOs. *See* Nongovernmental organizations

NIH. *See* National Institutes of Health
Nongovernmental organizations (NGOs), 20, 25
Nontarget organisms, effects on, 107–108
Nontarget species, effects on, 52
Nontransgenic male sterility, 76–77
Nontransgenic scions on transgenic rootstock, 83–84
  strengths, 83
  weaknesses, 83–84

O

Office of Management and Budget, 61
Oliver, Melvin, 23
Operational considerations, for bioconfinement, 180–198
Options for bioconfinement of plants, 110–113
  based on technology and gene-specific compounds, 185
  fitness handicaps, 112
  gene silencing, 111–112
  plastid engineering, 113
  sterility, 110–111
  tissue-specific expression, 112–113
  triploidy, 111
Organism choice, 96

P

Permits approved by APHIS for field tests in the United States
  genetically engineered turfgrass, 116–117
  genetically engineered turfgrasses, 116–117
  genetically engineered woody plants, 99–102
Phenotypic handicapping, 91–93, 169–173
  bacteria, 170–172
  fungi, 172–173
  strengths, 92
  viruses, 169–170
  weaknesses, 92–93
Physical barriers, 16
Physiochemical barriers, 16–17
Plant-made pharmaceuticals (PMPs), 24
Plants
  bioconfinement of, 65–129
  methods of bioconfinement, 4

Plastid engineering, 113
Pleiotropy, 183
PMPs. *See* Plant-made pharmaceuticals
Pollen-specific gene expression, 95
Population effects, 126–129, 177–178
Post-market Oversight of Biotech Foods, 35
Potential effects of bioconfinement, current and future GEOs, 36–38
Potential for gene flow, 118–120
Predation, heightened, 141
Private litigation
    decision-making about bioconfinement, 63–64
    and the efficacy of confinement, 193
Production cycle for all-female lines of fish, in species with an XY sex determination system, 143
Production site characteristics, 144–145
    strengths and weaknesses, 145
Programmed cell death (PCD), 75
Proposed bioconfinement of transgenic Atlantic salmon, 137–139
Proposed transgenic bioconfinement methods in plants, 73
Pseudomonas syringae, 22

R

RAC. *See* Recombinant DNA Advisory Committee
RAFI. *See* Rural Advancement Foundation International
Rationale for bioconfinement, 3–4
rDNA. *See* Recombinant DNA organisms
Recombinant DNA Advisory Committee (RAC), 21–22
Recombinant DNA (rDNA) organisms, 19–20
Recommendations
    to assess bioconfinement techniques with reference to temporal and spatial scales of field release, 184
    to assess the efficacy of bioconfinement, 196–197
    to assess the environmental effects of confinement failure, 196
    to compare the novel genotype with its progenitor before field release, 183–184
    to consider potential effects of a confinement failure on other nations, 194
    to consider the need for bioconfinement early in the development of a GEO or its products, 182
    to define an adequate level of bioconfinement early in the development of a GEO, 184
    to develop better understanding of invasion biology, 197–198
    to develop better understanding of the dispersal biology of organisms targeted for genetic engineering and release, 197
    to develop easily identifiable markers, 187–189
    to develop reliable, safe, and environmentally sound bioconfinement, 196
    to evaluate each GEO separately, 181
    to identify economic, legal, ethical, and social factors influencing application and regulation of particular techniques, 197
    to incorporate transparency and public participation in bioconfinement techniques and approaches, 189–190
    to pursue international cooperation to adequately manage confinement of GEOs, 194
    to support additional scientific research, 196–198
    to take human error into account when determining bioconfinement methods and evaluating their efficacy, 192
    to test confinement techniques before putting them into application, 183
    to use an integrated confinement system approach for GEOs that warrant confinement, 187
Recoverable block of function (RBF) technique, 112
Reducing exposure to transgenic traits, 93–96
    flower- and fruit-specific gene expression, 94–95

green-specific (chloroplast-targeting) gene expression, 93
pollen-specific gene expression, 95
roots and tuber-specific gene expression, 93–94
seed-specific gene expression, 95–96
vascular-tissue-specific gene expression, 94
Reducing gene flow to crop relatives, 85–93
chromosomal locations in allopolyploids, 88–90
cross-incompatibility, 88
fitness reduction in transgenic crop-wild progeny, 90–91
phenotypic and fitness handicaps, 91–93
repressible seed-lethal bioconfinement, 86
repressible seed lethal confinement, 85–87
Redundancy, 182–183
Reliable bioconfinement, recommendation to support additional scientific research to develop, 196
Repressible seed-lethal confinement, 85–87
strengths, 87
weaknesses, 87
"Repressor" molecules, 147
Resistance, delaying the evolution of, 52–53
Reversible transgenic sterility, 72–75
strengths, 74
weaknesses, 74–75
Risk assessment and management, 33
Risk assessment matrix, 32
Risks, 30–35
associated with gene flow into natural populations, 106–107
confinement failure with StarLink corn, 34
systematic risk assessment and management, 33
Risks of most concern with trees, 106–110
effects on nontarget organisms, 107–108
instability of transgene expression, 110
secondary phenotypic effects of transgenesis, 108–109
RNA silencing, 111–112
Rockefeller Foundation, 26
Rural Advancement Foundation International (RAFI), 25, 28

S

Safe bioconfinement, recommendation to support additional scientific research to develop, 196
Scale, changes of efficacy with, 184
Secondary phenotypic effects of transgenesis, 108–109
Seed saving, 26
Seed-specific gene expression, 95–96
strengths, 95
weaknesses, 95–96
Sexual reproduction
combining triploid sterilization with all-female lines, 142–144
disruption of, 132–145
production cycle for all-female lines of fish in species with an XY sex determination system, 143
production site characteristics, 144–145
sterilization through induction of triploidy, 133–141
in wild relatives, 140–141
Shellfish, bioconfinement of, 132–153
SIT. See Sterile insect technique (SIT)
Social acceptability of bioconfinement methods, 25–28
case study of the technology protection system— "terminator," 25–27
consequentialism and public acceptance, 27–28
Social factors influencing application and regulation of particular techniques, 197
Spatial scales
effectiveness of, 122–123
increases in and the efficacy of confinement, 193
"Split registrations," 34
Stability of transgenic confinement, 102–103
StarLink: Impacts on the United States Corn Market and World Trade, 35
StarLink corn, 23
StarLink™ Situation, The, 35
"Sterile feral" technology, 146
Sterile insect technique (SIT), 153–156
strengths, 156
weaknesses, 156

Sterile triploids, 23, 70–71
  strengths, 70
  weaknesses, 71
Sterility, 68–71, 110–111
  interspecific hybrids, 69
  unisexual plants lacking mates, 71
Sterilization through induction of triploidy,
      133–141
  normal steps in gamete fertilization
      and early cell division, 134
  proposed bioconfinement of
      transgenic Atlantic salmon, 137–
      139
  strengths, 133–135
  use of tetraploids to maximize triploid
      percentage, 136–141
  weaknesses, 135–136
Strategic public investment, in
      bioconfinement research, 195–
      198
"Suicide genes," 23, 173–176
  bacteria, 173–174
  fungi, 174–175
  viruses, 175–176
Superfund cleanup sites, 173

T

T-GURT. See Trait genetic use restriction
      technology
Taco Bell brand taco shells, 24
"Tandem constructs," 90
Technology Protection System (TPS), 23
Temporal scale
  effectiveness of, 124
  increases in and the efficacy of
      confinement, 193
"Terminator technology," 25–26
Tetraploids, used to maximize triploid
      percentage, 136–141
Tissue-specific expression, 112–113
TM. See Transgenic mitigation
TPS. See Technology Protection System
Trait genetic use restriction technology (T-
      GURT), 72, 85
Transgene expression
  artificially induced, 84–85
  instability of, 110
Transgene loss, 102–103
Transgenes absent from seeds and pollen,
      83–84

excision of transgenes before
      reproduction, 84
  nontransgenic scions on transgenic
      rootstock, 83–84
Transgenes in chloroplast DNA, 79–81
Transgenic algae, 121–122
Transgenic grasses, 115–121
  genetically engineered turfgrasses,
      115–120
Transgenic male sterility, 77–78
  strengths, 78
  weaknesses, 78
Transgenic mitigation (TM), 90
Transgenic sterile insects, 156–157
  strengths and weaknesses, 157
Transgenic sterility, 71–76
Transparency and public participation, 10,
      189–190
  and the efficacy of confinement, 189–
      190
  recommendation to incorporate
      transparency and public
      participation in bioconfinement
      techniques and approaches, 189–
      190
Trees. See Genetically engineered trees
Triploid sterilization, combining with all-
      female lines, 142–144
Triploidization, combining with
      interspecific hybrids, 152
  strengths and weaknesses, 152
Triploidy, 111

U

Understanding Risk, 31
Unisexual plants lacking mates, 71
  strengths, 71
  weaknesses, 71
U.S. Department of Agriculture (USDA), 1–
      2, 23, 182
  Agricultural Research Service, 25
  Animal and Plant Health Inspection
      Service, 24, 36, 54, 56, 115
U.S. Environmental Protection Agency, 1,
      24, 53, 62
U.S. Fish and Wildlife Service, 62
U.S. Food and Drug Administration (FDA),
      1, 61, 137–139
USDA. See U.S. Department of Agriculture

V

V-GURT. *See* Variety genetic use restriction technology
Variable functional sterility, 140
Variety genetic use restriction technology (V-GURT), 72–75, 84, 87
Vascular-tissue-specific gene expression, 94
Viruses
  bioconfinement of, 169–179
  displacement of indigenous populations, 164
  horizontal genetic transfer into local populations, 166
  invasion into indigenous populations, 161
  phenotypic handicapping, 169–170
  suicide genes, 175–176

W

Weediness, concerns about, 49–50
Wild hybrid, *Festuca arundinacea* and *Lolium multiflorum* Lam., 119